AI Native Enterprise

The Leader's Guide to AI-Powered Business Transformation

Yi Zhou

ArgoLong Publishing

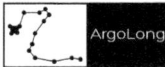

ISBN: 979-8-9893577-3-4 (hardback)

ISBN: 979-8-9893577-4-1 (paperback)

ISBN: 979-8-9893577-5-8 (eBook)

First edition 2024

For my mentors and followers, ignited by AI curiosity,

For Yan and Henry, my anchors of love,

And for everyone who has guided and stood by me,

This dedication is for you.

The Generative AI Revolution Series

Prompt Design Patterns: Mastering the Art and Science of Prompt Engineering (2023)

AI Native Enterprise: The Leader's Guide to AI-Powered Business Transformation (2024)

AI-Powered Next-Gen Workforce: The Comprehensive Guide to Microsoft 365 Copilot for Peak Performance (Coming soon)

Contents

AI-Native Thinking in Data Strategy and Data Management

Part Four: Embarking on Your AI Transformation Journey

Closing: Embarking on the AI Native Enterprise Journey

Preface

In an era where artificial intelligence (AI) is not just an advantage but a necessity, the journey to becoming an AI-native enterprise is both compelling and critical for leaders across industries. This book, "**AI Native Enterprise: The Leader's Guide to AI-Powered Business Transformation**", is born out of a convergence of my experiences, insights, and the urgent need I've observed in the business world for a comprehensive guide to harnessing AI's transformative power.

My inspiration to write this book stemmed from witnessing the struggle many organizations face in navigating the complexities of AI adoption and integration. The pace at which AI is evolving presents a unique set of challenges and opportunities. I realized that while many leaders recognize the importance of AI, there's a gap in understanding how to effectively transition to an AI-native enterprise. This book aims to bridge that gap, offering a roadmap for leaders to leverage AI for strategic advantage and transformative growth.

As you delve into the pages, you'll uncover the essence of what it means to be an AI-native enterprise. This book will guide you through the paradigm shift from conventional AI practices to an AI-first strategy, culminating in the adoption of an AI-native approach that reshapes your business DNA. It's designed to provide you with the insights and frameworks needed to navigate this journey, offering a blend of strategic imperatives, case studies, and practical tools that cater to both the seasoned AI practitioner and those new to the AI arena.

Overview of the Content

The book is meticulously structured into four pivotal parts, guiding leaders through the entire spectrum of AI integration in business:

1. **Generative AI Revolution and New Business DNA**: Part One explores the transformative impact of Generative AI (GenAI), charting the course from traditional AI applications to a strategic, AI-First approach, and ultimately to an AI-Native paradigm. This section aims to deepen leaders'

understanding of the GenAI revolution and its role in forging new business DNA.

2. **Creating a Compelling AI Vision and Winning Strategy**: Part Two provides a strategic framework for crafting an AI vision and strategy that paves the way for AI-native transformation. It introduces tools like the AI Business Value Octagon for strategic impact assessment, dives into financial analyses such as TCO and ROI, and proposes an AI Risk Mitigation Framework. This part offers actionable insights for incorporating AI into strategic planning and risk management, facilitating a successful shift to AI-Native enterprises.

3. **Implementing a Robust AI Operating Model**: The third part addresses the creation of a comprehensive AI Operating Model, focusing on workplace transformation, AI governance, and the essential technological and data foundations for AI integration. It outlines strategies for developing AI-ready talent and culture, ensuring responsible AI practices, and establishing a strong technology and data base, guiding leaders towards crafting organizations ready for AI-driven operations.

4. **Embarking on Your AI Transformation Journey**: Part Four details the initial steps and strategies for beginning the AI transformation journey, emphasizing the importance of prompt engineering and offering frameworks for a complete AI Native transformation. This section provides practical advice for overcoming early obstacles, mobilizing teams, and strategically embedding AI into business operations and culture, ultimately leading to successful AI adoption and transformation.

"AI Native Enterprise" is designed as a holistic manual, equipping leaders with essential knowledge, strategies, and tools for navigating the AI-powered business arena, from foundational concepts to strategic execution, ensuring AI-driven excellence across their organizations.

The benefits of embracing an AI-native strategy are immense, extending beyond operational efficiency to fundamentally redefining competitive advantage and customer experience. Through a detailed exploration of generative AI, strategic transformation, and the nuances of AI governance, this book aims to equip you with the knowledge to lead your organization into a future where AI is seamlessly integrated into every facet of your business.

My journey in compiling these insights has been one of discovery, challenge, and ultimately, profound learning. It is my hope that "AI Native Enterprise" serves not just as a guide but as a catalyst for innovation and transformation within your organization. I invite you to engage with the material, apply the strategies, and embark on the path to becoming an AI-native enterprise.

As we stand on the brink of a new era in business, the decision to embrace AI is not just strategic—it's existential. The future belongs to those who are prepared to lead the charge in this AI-powered revolution. Welcome to the journey.

Who Should Dive Into This Book?

"**AI Native Enterprise: The Leader's Guide to AI-Powered Business Transformation**" is crafted for a wide array of readers who are poised at the brink of integrating AI into their strategic vision or looking to deepen their understanding of AI's transformative power in the business landscape. Here's a closer look at who will benefit most from this comprehensive guide:

- **Executives and Board Members**: Senior leaders will find invaluable insights into steering their organizations through the complexities of AI adoption, ensuring alignment with long-term business goals and navigating the strategic challenges and opportunities presented by AI.

- **Business Leaders**: Those charged with charting the future course of their companies will gain deep insights into leveraging AI for competitive advantage, understanding emerging AI trends, and crafting strategies that embed AI at the core of business innovation.

- **Investors**: Savvy investors looking to understand where and how AI is creating new business opportunities and disrupting industries will find critical analyses and forecasts that can guide investment decisions in technology and AI-centric companies.

- **Technical Leaders**: CTOs, CIOs, and other technical executives responsible for the implementation and governance of AI technologies will appreciate the deep dive into AI operating models, governance frameworks, and the technological and data foundations essential for AI integration.

- **Business Strategists**: Professionals focused on innovation, growth, and competitive strategy will find frameworks and models to help think through how AI can be used to create new value propositions and business models.

- **Transformation Leaders and Change Agents**: Those leading the charge on organizational transformation will gain insights into managing change, cultivating an AI-ready culture, and aligning organizational structures and talent with an AI-driven future.

- **AI Professionals and Trailblazers**: Practitioners and pioneers in the field of AI will discover advanced discussions on prompt engineering, AI ethics, and the latest in AI research and applications, providing a rich source of inspiration for their work.

- **Academic Luminaries and Scholars-in-the-Making**: Educators and students in business, technology, and AI disciplines will find the book a valuable addition to their curriculum, offering a comprehensive view of AI's impact on business and society along with forward-looking perspectives on AI innovation.

- **Anyone Interested in AI-Powered Business Transformation**: Curious minds eager to understand the potential of AI to redefine the business landscape will find this book an accessible yet thorough guide to the principles, strategies, and practices that underpin the AI-native enterprise.

Whether you're at the helm of an organization, pioneering new AI technologies, investing in the future of business, or simply fascinated by the intersection of AI and commerce, "AI Native Enterprise" provides the insights, strategies, and knowledge you need to navigate the exciting terrain of AI-powered business transformation.

Introduction

"Intelligence is the ability to adapt to change." —— *Stephen Hawking*

I n the unfolding narrative of human progress, we are witnessing the ascent of artificial intelligence—a revolution that's reshaping the very fabric of society. This isn't a distant phenomenon; it's a current reality that's already transforming our professional lives, streamlining communication, and enhancing creativity.

AI has transitioned from the abstract into a practical force that shapes our digital interactions, listens and responds to our commands, and innovates in art, music, and writing. It's the powerful analyst that deciphers vast data arrays to reveal trends and insights that escape human detection. AI represents the future—a future where technology doesn't just follow orders but foresees our needs, relates to our emotions, and expands our capabilities.

As leaders in this new era of AI, we're not just observers; we're shaping its impact. Our strategies, our questions, and our aspirations will steer this technological evolution. Let's approach this era with the foresight, readiness, and vision befitting our roles, for we're not just witnessing a change—we're leading it.

The Evolution of AI Fields

1. Artificial Intelligence (AI)

> **Definition:** Artificial Intelligence (AI) refers to the capability of a machine to imitate intelligent human behavior. AI systems can perform tasks such as learning, reasoning, perception, problem-solving and decision making.

The vision of developing thinking machines and AI was introduced in the mid 1900s. In 1950, Alan Turing published a seminal paper entitled "Computing Machinery and Intelligence" which proposed a test for machine intelligence now known as the Turing Test. This sparked debate around the possibility of machines rivaling human cognition.

Other pioneering early work exploring concepts adjacent to AI includes McCulloch and Pitts' 1943 research modeling artificial neurons and Donald Hebb's 1949 book "The Organization of Behavior" discussing neural networks and learning. In the 1950s, scientists also began creating programs capable of solving algebra, chess and logic problems, demonstrating rudimentary reasoning skills.

The 1956 Dartmouth Conference hosted by John McCarthy, Marvin Minsky and others then officially birthed AI as a field and community devoted to formally pursuing this vision with multidisciplinary perspectives. Attendees proposed grand ideas about replicating aspects of human intelligence in machines. This catalyzed significant funding and support for early AI labs at institutions like Stanford and MIT during the 1960s.

Yet by the 1970s, disillusionment arose from difficulties realizing some initial lofty goals as the limitations in areas like knowledge representation emerged. Funding declined as a result during this "AI winter" period. Still, critical progress continued in fields like expert systems, natural language processing and machine learning laying foundation for the future revitalization of AI innovation.

2. Machine Learning (ML)

Definition: Machine learning refers to the study of computer algorithms that improve automatically through experience and by the use of data. It is seen as a subset of artificial intelligence. Machine learning algorithms build a mathematical model based on sample data, known as "training data", in order to make predictions or decisions without being explicitly programmed to perform the task.

Machine Learning (ML) emerged in the late 1950s as an academic discipline focused on making computers learn from data without explicit programming. In 1959, IBM researcher Arthur Samuel coined the term "machine learning" and developed programs for checkers that could learn from experience. This demonstrated principles of having computer programs iteratively improve through data over time.

Some consider the 1952 paper by Arthur Samuel on checkers-playing programs as the genesis of machine learning as a subfield distinct from earlier neural network and related research. Others point to Frank Rosenblatt's 1957 work on the perceptron algorithm for pattern recognition as pivotal in spurring interest in machine learning specifically. By 1960s, the field had advanced to computers recognizing shapes, faces and cursive writing samples.

Although the concepts and some applications took root in the the 1950-60s pioneers, machine learning gained significantly increased mainstream interest starting in the 1990s and accelerating further in the 2000s. This resurgence resulted from a confluence of advancements in computational statistical models and algorithms, growth of available training data with the rise of the internet and "big data", and access to enhanced parallel computational power through improvements in computer chips and GPU processing.

With these improved capacities, practical deployments of machine learning systems for real-world applications in diverse areas blossomed. Industry adoption grew more ubiquitous through the 2010s with major leaps in accuracy and capability of supervised, unsupervised, semi-supervised and reinforcement learning models across tasks like image classification, machine translation, anomaly detection and game strategy. Excitement now abounds about future machine learning possibilities.

3. Deep Learning (DL)

Definition: Deep learning refers to artificial neural networks composed of multiple layers that can learn hierarchical representations of data. Deep learning algorithms perform successive layers of abstraction and representation learning to discover complex patterns from raw data.

The origins of deep learning trace back to the 1940s and the early neural network pioneers. Neuroscientist Warren McCulloch and mathematician Walter Pits modeled the first artificial neurons in 1943. In 1949, Donald Hebb presented theories of neural plasticity and learning. These conceptual foundations influenced later researchers in conceiving algorithms mimicking biological neural networks.

In the 1950s, early incarnations of multi-layer networks start emerging. In 1965, Alexey Ivakhnenko and Lapa published the first working deep learning algorithm showing superior human handwriting recognition compared to single layer models. By 1974, seminal work by Paul Werbos introduced the backpropagation algorithm vital for training deep neural networks.

By the 1980s, key deep learning methods like convolutional neural networks (CNNs) and long-term short memory networks (LSTMs) emerged. But long training times and computational limitations obstructed progress. The concepts surpassed practical application capacities of the time. This changed in 2006 and 2012 when papers conclusively showed deep learning models surpassing past benchmarks by large margins in landmark tasks like computer vision and automatic speech recognition.

This success resulted from newly applied methods combining improved neural network architectures with increases in training data and advancements in computational power, especially graphical processing units (GPUs). Starting in 2009 and the early 2010s, GPUs enabled training complex models with millions of parameters on large datasets in feasible times. With this expanded capacity, deep learning adoption grew exponentially across domains throughout the 2010s and continue today as the state-of-the-art approach for many AI problems.

4. Natural Language Processing (NLP)

> **Definition:** Natural language processing (NLP) refers to the ability of a computer program to understand, interpret, and manipulate human language. NLP enables computers to perform useful tasks with natural language including translation, sentiment analysis, speech recognition, and text generation.

The origins of NLP date back to Alan Turing in 1950 publishing his seminal paper "Computing Machinery and Intelligence" which introduced the Turing test as a metric for intelligence. This catalyzed interest around machines processing and producing human language.

In the 1950s and 1960s, early NLP work focused on machine translation by creating rules-based systems to convert text between languages. Progress was slow due to the complexity of human language. The Georgetown experiment in 1954 on fully automatic translation began to expose these challenges.

In the 1960s and 1970s various new approaches emerged. Chomskyan linguistics introduced ideas like context-free and transformational grammars. SHRDLU in 1968 demonstrated simple English language understanding and generation. These traditions continued advancing core NLP tasks like part-of-speech tagging and parsing through the 1980s.

The rise of machine learning in the 1980s and statistical NLP in the 1990s marked a major transition, allowing more data-driven versus strictly rules-based NLP models. By 2000s, statistical machine translation became dominant. More recently, the resurgence of neural networks and deep learning combined with enormous data accelerated NLP capabilities even further, achieving new state-of-the-art results in machine translation, dialogue systems, and language generation. NLP has become ubiquitous in applications today due to these advances.

5. Predictive AI

Definition: Predictive AI refers to artificial intelligence systems focused on making predictions about future events or outcomes based on historical data. It involves analyzing past data sets, identifying patterns, developing a model correlating variables to an outcome of interest, and using that model to forecast what might happen in new scenarios.

The desire to predict the future is ancient, but applying AI capabilities towards predictive modeling began in the 1950s. Herbert Simon and Allen Newell developed the first predictive program in 1957 called the General Problem Solver, one of the original AI programs. In the 1960s and 70s, Expert systems aimed to make predictions and recommendations by attempting to mimic human decision-making.

These early promising works faced challenges in handling complexity and uncertainty. The mainstream adoption of predictive AI required expanded data and computing, especially increases in data storage and processing that emerged in the 1990s and 2000s. With larger datasets combined and more advanced algorithms like machine learning to discern signals, predictive analytics gained significant commercial use.

Increasing real-time data now allows prediction systems to dynamically incorporate new information. Deep learning further elevated predictive intelligence by discovering subtle correlations. Cloud computing also enabled scaling the infrastructure to support vast predictive models. The applications of predictive AI now span across sectors like healthcare, finance, transportation, and more. Accuracy and decision support capabilities will continue advancing into the future.

6. Generative AI

> **Definition:** Generative AI (GenAI) refers to a type of artificial intelligence that can create new artifacts, such as text/document, images, audio/video, idea/design, software code, strategy or method, by learning from existing data, allowing it to generate novel, realistic outputs that didn't previously exist. Notable examples include OpenAI's ChatGPT (GPT-4 Model) and Google's Bard (Gemini Model).

Early work exploring procedural content generation with computers dates back to the 80s and 90s in areas like computer graphics. But recent exponential progress was catalyzed by the rise of deep learning and related methods.

A breakthrough work was Ian Goodfellow's 2014 paper introducing generative adversarial networks (GANs). This architectural paradigm set up an adversarial game between two neural networks - one generating candidates and one evaluating realism - to rapidly enhance synthetic outputs. Enhancements to GANs, variational autoencoders (VAEs), and diffusion models soon significantly elevated generative capacity and quality across modalities.

By the late 2010s, generative models were producing remarkably realistic and high-resolution synthetic images, audio, video and text. The applications span areas like content creation to drug design to autonomous sensor simulation for robotics systems.

Generative AI gained significant recent mainstream attention thanks to systems like ChatGPT launched in late 2022, which can generate human-like text responses to natural language prompts. The public interest around such technologies highlights the accelerating progression of generative models' capabilities. Future frontiers include interactive human-AI co-creation and increased customization control over synthetic data parameters and attributes.

Connected Evolution of AI

The evolution of artificial intelligence has been marked by meaningful interplay between its subfields, where progress in one area contributes to advancements in others. The symbiotic growth reinforcing the collective progress is a hallmark of AI.

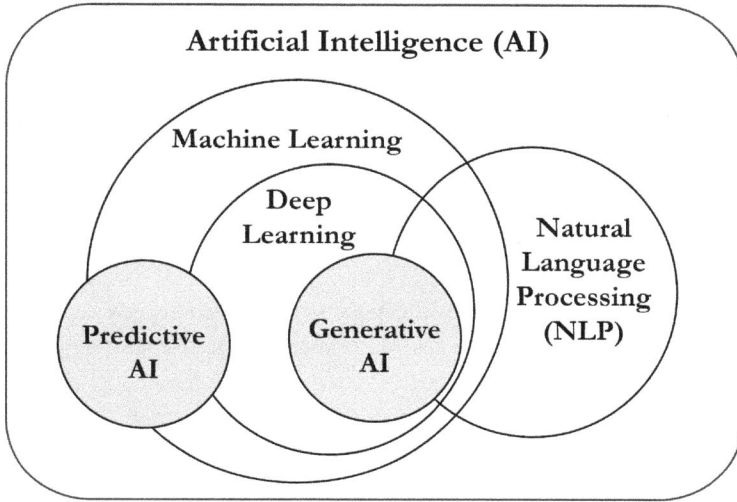

Figure 1: The Relationships of Key AI Components

In the early decades, research between neural networks, machine learning, natural language processing and adjacent fields influenced each other. As an example, machine learning helped overcome limitations of rules-based systems in NLP during the 1980s and 1990s transition toward statistical NLP.

More recently, breakthroughs in deep learning ignited new capabilities by creating more advanced neural network architectures. For instance, the rise of deep learning-based translation models and word embeddings enriched many downstream NLP tasks.

Similarly, generative adversarial networks originating from deep learning paradigms enhanced creative generation across areas like image, video, audio and text synthesis. Improvements in natural language processing have also provided building blocks powering the natural language generation capacities seen in systems like ChatGPT.

On the predictive side, stronger machine learning approaches have enabled identifying subtle signals in big data to fuel predictive analytics. And generative modeling advancements have allowed creating high-fidelity simulated data to train predictive models for scenarios lacking historical examples.

The interwoven acceleration also relates to shared infrastructure improvements providing multiplier effects. GPU computing, curated datasets, increased model scale, streaming data systems, open source frameworks and cloud-based tooling gave a common boost to otherwise disconnected domains in AI.

Emerging trends continue to highlight this interdependence. Self-supervised learning advancements are transferring across computer vision, NLP and beyond. Reinforcement learning progress informs advances in areas like robotics and protein folding. As the collective field advances, the symbiotic relationships between subfields persist as a catalyst for future innovations.

Understanding Generative AI

Generative AI represents a significant leap forward in artificial intelligence, offering revolutionary applications in various fields such as natural language processing, image generation, and more. At its core, Generative AI refers to algorithms capable of generating new, unique content or data that resemble human-like creativity and understanding. This section delves into foundational concepts integral to Generative AI, providing clear definitions and insights into foundation models, large language models (LLMs), and ChatGPT.

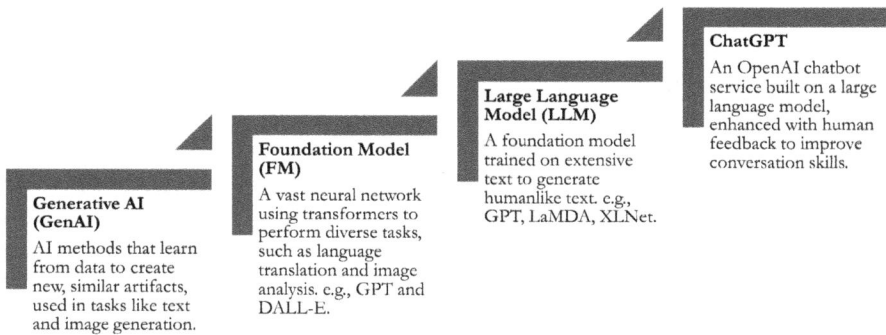

ChatGPT

An OpenAI chatbot service built on a large language model, enhanced with human feedback to improve conversation skills.

Large Language Model (LLM)

A foundation model trained on extensive text to generate humanlike text. e.g., GPT, LaMDA, XLNet.

Foundation Model (FM)

A vast neural network using transformers to perform diverse tasks, such as language translation and image analysis. e.g., GPT and DALL-E.

Generative AI (GenAI)

AI methods that learn from data to create new, similar artifacts, used in tasks like text and image generation.

Figure 2: Key Concepts in Generative AI

Foundation Models (FMs)

A foundation model is a broad and versatile machine learning model trained on a vast and diverse dataset. These models are foundational in two key aspects: first, in their ability to be fine-tuned for a wide range of tasks beyond those they were initially trained for; and second, in serving as a base for developing more specialized models. The training process, involving extensive and diverse datasets, endows these models with a rich understanding of patterns, language, and even visual concepts. This versatility makes foundation models a cornerstone in the landscape of Generative AI.

Large Language Models (LLMs)

Large Language Models (LLMs) are a subset of foundation models specifically designed for understanding, generating, and interacting with human language. Characterized by their vast number of parameters (often in the billions), these models are trained on extensive corpuses of text data. This training enables them to perform a wide array of language-related tasks such as translation, question-answering, and text generation with remarkable proficiency. LLMs are adept at understanding context, nuances, and even the subtleties of different languages and dialects, making them powerful tools in the realm of natural language processing.

ChatGPT: The Conversational Milestone

ChatGPT, a specific instance of a large language model, is designed for generating human-like text in a conversational context. Developed by OpenAI, it stands out for its ability to engage in dialogues, answer questions, and provide information or creative content in a manner that closely mimics human conversation. ChatGPT is built upon the GPT (Generative Pre-trained Transformer) architecture, which allows it to generate coherent and contextually relevant responses. Its training involves not only a vast dataset of text but also reinforcement learning techniques to refine its responses based on human feedback, ensuring a high level of relevance and appropriateness in its interactions.

The realms of foundation models, large language models, and specific implementations like ChatGPT represent the cutting edge of Generative AI. Their capabilities extend far beyond simple data processing, venturing into the domain of creativity and complex problem-solving. As these technologies continue to evolve, they promise to unlock new potential and applications, fundamentally altering our interaction with machines and digital content.

GPT-4: A New Era of Intelligence

The Microsoft Research team's "Sparks of Artificial General Intelligence: Early Experiments with GPT-4" presents an in-depth analysis of OpenAI's GPT-4. This 155-page paper posits GPT-4 as a stepping stone toward **Artificial General Intelligence** (AGI), given its broad-spectrum capabilities. GPT-4's prowess in language, coding, visual tasks, and interdisciplinary problem-solving highlights its potential as an AGI harbinger.

Notably, GPT-4's linguistic acumen shines, crafting responses with a clarity and context that surpass its predecessors like GPT-3. Its programming intelligence is equally impressive, generating complex animations and drafting visual representations in code, marking a leap in AI's coding applications.

Surprisingly, GPT-4 excels in vision tasks, capable of creating visual code for graphics like unicorns, showcasing its multifaceted nature. This interdisciplinary intelligence, spanning domains from medicine to law, underlines its comprehensive problem-solving approach.

GPT-4 also exhibits a nuanced understanding of human emotions, a critical aspect of social intelligence, vital for interactive AI applications. Yet, its limitations in mathematics and goal-oriented tasks point to the need for evolution beyond next-word prediction to realize full AGI.

The journey toward AGI is fraught with hurdles, but GPT-4's emergence, alongside models like Google Gemini, illustrates significant strides in AI advancement.

Generative AI vs. Traditional AI

Artificial intelligence (AI) has advanced rapidly in recent years into two broad categories - traditional AI and generative AI. While both leverage machine learning and neural networks to enable computers to perform human-like tasks, there are some key differences between the two AI types.

Traditional AI, also referred to as predictive or narrow AI, focuses on analyzing data to uncover patterns and insights. It then uses these patterns to make numeric predictions and recommendations for specialized use cases. For example, predicting which customers are likely to churn, forecasting product demand, detecting fraud, personalizing content recommendations, and optimizing manufacturing quality.

Generative AI refers to a newer class of AI systems with the ability to generate completely novel content based on their contextual understanding gained from ingesting massive amounts of data. This includes generating text, code, images, audio, video, and synthetic data that are often indistinguishable from human-created content. Generative AI can be applied to broader use cases beyond numeric predictions.

The table below summarizes some of the key differences between traditional and generative AI:

Parameter	Traditional AI	Generative AI
Objective	Analyze data, uncover patterns and insights to make predictions	Generate new content and assets based on contextual understanding
Data Used for Training	Carefully curated, domain-specific data	Publicly available data from the internet
AI Models	Narrowly focused on specialized use cases	Foundation models trained on massive datasets
Applications	Numeric predictions for specific business issues	Broad content creation capabilities
Ease of Use	Requires AI and domain expertise	More accessible to general users
Output Type	Recommendations, predictions	Text, images, audio, video, synthetic data

Table 1: The Comparison of Traditional AI and Generative AI

Some examples help illustrate how generative AI differs from traditional AI:

- A manufacturing firm uses computer vision algorithms to automatically scan product images on the assembly line. This traditional AI system flags defects and deviations from quality norms. A generative AI system could generate additional synthetic images of products to augment the training data.

- A bank uses ML models to analyze account transaction patterns and predict potential fraud. A generative AI chatbot could respond to customer inquiries about new account openings and transaction disputes.

- A university enrollment department uses predictive models to forecast application volume. A generative AI tool helps admissions counselors draft personalized email campaigns and website content tailored to prospective student profiles.

As these examples demonstrate, while traditional AI focuses on numeric predictions, generative AI expands the possibilities for automating content creation and complex human tasks. The two complement each other, with traditional AI continuing to play a key role in analyzing data for focused domains and generative AI providing a user-friendly interface for broader applications. As AI capabilities grow more advanced, harnessing both predictive and generative models will become critical for businesses seeking a competitive edge.

The Best of Both Worlds: Hybrid AI Solutions

Individually, traditional AI and generative AI bring valuable yet distinct capabilities. However, when used in combination, they can achieve remarkable results not possible by either in isolation.

For example, in software development, traditional AI can analyze source code to flag vulnerabilities and suggest fixes. But to rapidly develop secure new code, generative AI can synthesize entire functional components aligned to specifications. Together these accelerate releasing high-quality software.

In manufacturing, computer vision inspection systems (traditional AI) ensure quality control and detect defects. When defects arise, generative AI can propose design tweaks based on its broad understanding. This boosts yield rates. Further, generative AI can create photorealistic synthetic images to expand training data and improve model accuracy.

For customer support, virtual assistants use natural language processing (traditional AI) to understand and respond to inquiries. To handle more complex issues, conversational agents can pass the context to a generative AI assistant that provides thoughtful, empathetic responses indistinguishable from human agents.

In pharmaceutical research, predictive analytics uncover patterns in patient data to inform clinical trial participant selection and recruitment. Meanwhile, generative AI synthesizes additional anonymized patient health records protecting privacy while accelerating research.

As these examples illustrate, combining solutions creates symbiotic systems exceeding individual capabilities. Traditional AI handles analysis for defined use cases. Generative AI brings broad knowledge allowing fluid conversations and creative adaptations suiting novel situations. Together they will drive the next wave of AI innovation.

Beyond Imagination: The Transformative Power of GenAI

The advent of generative AI (GenAI) represents a seismic shift comparable to the biggest technological upheavals in history. Experts have likened the breakthroughs in models like DALL-E 2, ChatGPT and others to the disruption sparked by transformative innovations such as the steam engine during the first industrial revolution. The implications span far and wide, impacting everything from startup innovation velocity to new paradigms for human-computer collaboration.

Generative AI: The Architect of Tomorrow's Industries

In the ever-evolving tableau of technological advancement, Generative AI emerges as the linchpin of innovation, redefining and reshaping industry, science, and daily life with its versatile prowess. Let's embark on an exploratory journey through the multifaceted applications of Generative AI:

Reinventing Business Dynamics: Generative AI is not just altering the business ecosystem; it's recreating it. By leveraging its transformative potential, businesses are discovering new horizons of efficiency and customization.

Evolving Customer Relations: Imagine customer interactions empowered by AI that doesn't just react but anticipates needs. Generative AI infuses customer engagements with a mix of personalized efficiency, setting a new standard for consumer experience.

The New Age of Product Development: The fusion of human creativity with the analytical power of Generative AI is revolutionizing product design. It pioneers a movement of intuitive and consumer-aligned innovations, transforming imagination into tangible reality.

Revolutionizing Supply Chain Management: Generative AI enhances supply chains beyond their logistical functions, imbuing them with predictive analytics and adaptive strategies that redefine the concept of supply and demand logistics.

Crafting Tailored Industry Solutions: The chameleon-like versatility of Generative AI is invaluable in devising specialized solutions that cater to the unique challenges and demands of various industries.

Strategic Business Planning: Generative AI transcends traditional strategic planning. It simulates and predicts market dynamics, equipping businesses with strategies that are both resilient and visionary.

Financial Foresight: In the financial sector's complex labyrinth, Generative AI emerges as a clarifying force, offering acute investment insights and market forecasts that equip financial professionals with unparalleled strategic acumen.

Innovations in Biopharma: Generative AI stands at the forefront of biopharmaceutical research, accelerating the journey from laboratory to treatment with its capacity to analyze complex datasets, potentially transforming healthcare innovation.

Enriching Personal Spaces: Generative AI brings a personalized touch to domestic living, tailoring everything from meal plans to interior décor, ensuring our homes reflect and adapt to our evolving preferences.

Personal Productivity Enhancement: As an invisible ally, Generative AI optimizes personal productivity. It harmonizes schedules, curates learning resources, and anticipates burnout, promoting a balance between efficiency and wellness.

Personal Development Journey: Generative AI crafts customized learning experiences, aligning ambitions with tailored resources and pacing, making personal growth an intimate and impactful adventure.

Tailored Travel Experiences: Generative AI weaves together personal tastes, past travels, and global trends to design travel experiences that are not just trips but transformative journeys.

As we venture into an era where Generative AI integrates into the very essence of our lives, the scenarios envisioned here are but a glimpse into the vast expanse of its transformative impact. Embracing this technology paves the way for a future rich with unprecedented innovations and boundless possibilities.

Rapid Disruption

Whereas previously it may have required thousands of employees for a startup to disrupt entrenched incumbents, GenAI profoundly democratizes capabilities allowing small teams to take on giants. For instance, Midjourney generated unicorn status and a billion-dollar valuation with only 11 staffers by leveraging AI image generation. These models can automate complex tasks that previously required specialized skills, resources or labor. Startups embracing GenAI can unlock unprecedented productivity, creativity and problem-solving capacity.

The bandwidth of human attention, resources and effort impose inherent constraints to growth velocity for conventional businesses. In contrast, AI-based startups can scale impact and deliver value at speeds exponentially greater than human-only organizations. Cloud infrastructure further enables launching and iterating AI models without massive capital outlays upfront.

Once the AI is trained, each additional query contributes marginal compute costs yet yields brand new creative output tailored to prompt requests. This asymmetric returns equation fundamentally alters the relationship between inputs and outputs. Whereas linear growth followed conventional assumptions, AI opens frontier potential for non-linear, exponential gains as models ingest more data and users to improve.

Early examples foreshadow coming waves of GenAI-powered disruption across sectors. In biomedicine, Anthropic claims to have reduced drug discovery timelines by over 95% using AI. New AI art houses like Cabinet NOW! sell millions in inventory within months. Soon AI may augment everything from content creation to chip design and beyond. Incumbents wed to "BC" ways face mounting threats from fledgling insurgents pushing boundaries with AI.

New Modes of Thinking

Successfully navigating the GenAI age necessitates reorienting mindsets attuned to pre-ChatGPT eras. With these powerful tools, past experience may prove obsolete for solving contemporary challenges. Leading organizations openly acknowledge the need to abandon prior assumptions and solutions ingrained before recent AI progress in order to pave the way for cutting-edge innovations. Failure to rewrite mental models for the present reality risks getting left behind.

In a 2022 world, certain tasks would be assumed to require specialized expertise, extensive resourcing or complex coordination. But GenAI upends these conventions. For example, an animation studio previously needed entire teams of artists to storyboard scenes. Today, a prompt suffices for AI to generate detailed visualizations.

Similarly, certain research endeavors may have demanded months of work pre-ChatGPT. Enterprises relied on intuition honed over years to guide strategy. Now AI handles literature reviews in seconds while also providing creative direction by assessing vast datasets. Even complex legal and medical roles may transform with tools capable of parsing details human specialists cannot.

Rather than doubling down on traditional workflows, visionary leaders openly question every assumption in light of new capabilities. They vigilantly guard against creeping biases rooted in outdated perspectives. Existing processes that worked well in 2022 may actually obstruct progress today. Shedding limitations of the past proves essential to fully activate GenAI's future.

Co-Creation With AI

Whereas past technological advances often replaced human input, GenAI uniquely enables collaboration between human and artificial intelligence - combining strengths while mitigating weaknesses. Researchers describe this as making humans "better than the sum of their parts." Effective integration of subject matter experts alongside models like ChatGPT for research, content creation, data analysis and more exemplify emerging paradigms for hybrid intelligence. Those who learn to leverage symbiotic relationships with AI will excel.

GenAI excels at tasks like consuming or generating vast information, rapid calculation, and tireless iteration. Yet it currently struggles with context, causality, empathy and other facets innate to human cognition. On the other hand, people find open-ended abstraction, infinite recall and high-volume output difficult whereas AI thrives on these fronts. This presents prime opportunities for mutual improvement.

Pioneering organizations actively facilitate humans and AI peer-level collaborations. Writers partner with language models to co-author stories, textbook chapters and analysis together. Scientists integrate AI assistants into experimental protocols for collecting, processing and reviewing relevant literature. Product developers use generative design to accelerate prototyping then refine options selected by human judgment.

Rather than a zero-sum relationship, constructive human-AI symbiosis paves the path toward previously unfathomable creative potential being realized. A future filled with groundbreaking discoveries, inventions and art shaped by this cooperative dynamic appears imminent as the GenAI revolution continues unfolding. The wise will waste no time learning to tap into these collaborative superpowers.

Once-In-A-Life Opportunity

The societal transitions spurred by steam, electricity and computers unfolded over decades. In contrast, GenAI's immense power leaves little time for gradual adjustment. Organizations must rapidly reinvent to keep pace with accelerating progress. How they respond now may determine competitive viability through the coming age of artificial general intelligence.

The current state of generative AI already far surpasses expectations projected just a couple years ago. In the blink of an eye, fathomless applications materialized at blinding speed. Extrapolating the exponential momentum hints at even more profound capabilities arriving faster than imagined.

Unlike past infrastructural transformations, individuals may only get one shot in life at witnessing - and harnessing - the radical metamorphosis of labor and technology underpinning modern civilization. The GenAI big bang constitutes such an epoch-defining milestone ripe with opportunity for those bold enough to seize the moment.

Visionaries across fields now face a watershed juncture. The choice? Embrace imaginative AI collaboration to unleash newfound creative powers - or cling to old ways and risk disruption into demise. For most, ignoring this once-in-a-generation calling looms unthinkable. Destiny awaits those who hear the call.

Part One: Generative AI Revolution and New Business DNA

ArgoLong Publishing

Chapter 1

Generative AI Revolution for Next-Gen Business

"If our era is the next Industrial Revolution, as many claim, AI is surely one of its driving forces." — Fei-Fei Li

As we embark on the third decade of the 21st century, the business landscape is witnessing a seismic shift brought about by the advent of Generative AI (GenAI). This technology, characterized by its ability to learn from data and generate new, original content and solutions, is not just an incremental advancement; it's a revolutionary force redefining the very fabric of how businesses operate, compete, and innovate.

GenAI stands at the intersection of machine learning, data analytics, and creative algorithmic prediction. Unlike traditional AI, which primarily focuses on analyzing and interpreting data, GenAI takes a step further by generating new content, ideas, and strategies. This includes everything from creating realistic images and writing code to proposing business strategies and inventing new products.

The journey to this point has been evolutionary. The initial phases of AI in business were about automation and efficiency—doing things faster and more accurately. Then came the era of Big Data and analytics, where the focus shifted to understanding and predicting trends. Now, GenAI is propelling us into a future where AI is not just a tool for processing and analyzing but an active participant in creation and innovation.

This new era marks a paradigm shift in business. GenAI introduces a level of dynamism, adaptability, and innovation previously unattainable. It's not just about using AI to improve existing processes; it's about using AI to create new processes, products, and ways of thinking. The implications are vast - from transforming product development and customer engagement to revolutionizing market strategies and operational models.

GenAI's impact is all-encompassing, cutting across various sectors and functions. In marketing, it's about creating more personalized and engaging content. In product development, it's about faster innovation cycles. In decision-making, it's about AI-driven strategies that can adapt to changing market dynamics in real-time.

With great power comes great responsibility. The rise of GenAI brings with it a host of challenges—ethical, regulatory, and operational. Businesses must navigate issues like data privacy, algorithmic bias, and the potential displacement of jobs with care and foresight. There's also the challenge of integrating GenAI into existing business structures in a way that complements rather than disrupts.

As business leaders, entrepreneurs, and innovators, understanding and harnessing the power of GenAI is no longer just an option; it's imperative for success in the ever-evolving business landscape. As we delve deeper into this chapter, we will explore how GenAI is shaping the future of business across various domains. We'll look at its applications, the emerging business models it's spawning, and the strategies companies are employing to leverage this technology for competitive advantage.

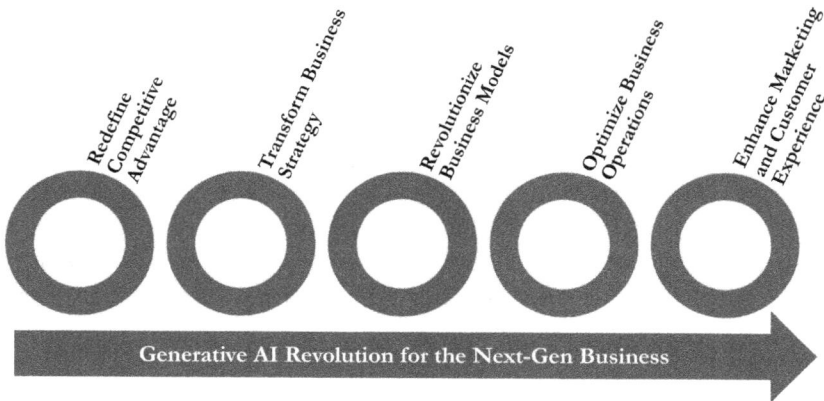

Figure 1-1: Top 5 Business Impact of Generative AI Revolution

Redefining Competitive Advantage

In the GenAI era, the landscape of competitive advantage is undergoing a radical transformation. Traditional pillars of competitiveness like cost efficiency, speed, and quality remain fundamental, but GenAI introduces a new set of dimensions that are reshaping what it means to have an edge in the marketplace.

- **Data Supremacy**: The advent of GenAI has elevated the importance of data to new heights. In this context, competitive advantage increasingly hinges on a company's ability to gather, process, and leverage data. The quality, diversity, and volume of data a business can access directly influence the effectiveness of its AI applications. Companies with rich data resources can train more sophisticated AI models, leading to more accurate predictions, insights, and innovations. This data supremacy is not just about the quantity of data but also about the depth and breadth of insights that can be extracted from it.

- **Algorithmic Efficiency**: The efficiency and sophistication of AI algorithms are now critical competitive factors. Businesses that develop or acquire superior AI models can gain significant advantages in terms of innovation, efficiency, and decision-making. These algorithms can identify patterns and insights that are imperceptible to human analysts, leading to breakthroughs in product development, market strategy, and operational efficiency. Algorithmic efficiency also encompasses the ability to continuously improve and adapt these models in response to new data and changing market conditions.

- **Adaptability and Speed**: GenAI endows businesses with the ability to rapidly adapt to market changes and consumer trends. This agility is a significant competitive advantage in today's fast-paced business environment. Companies leveraging GenAI can quickly pivot their strategies, operations, and even business models in response to emerging opportunities and threats. This adaptability is not just reactive but also predictive, as AI models can forecast market trends and shifts, allowing businesses to stay ahead of the curve.

- **Customization and Personalization**: GenAI enables businesses to offer unprecedented levels of customization and personalization in their products and services. This capability can be a strong differentiator in markets where consumers increasingly value experiences and products that are tailored to their individual preferences. Businesses that effectively utilize GenAI to understand and cater to these specific needs can build deeper customer relationships and loyalty.

- **Innovation as a Core Competency**: With GenAI, innovation becomes a continuous, data-driven process. Companies that can harness AI to consistently generate new ideas, solutions, and products are likely to outpace competitors. This constant stream of innovation, fueled by AI's ability to combine and recombine vast arrays of information, can lead to the development of novel products and services, opening up new markets and opportunities.

- **Ethical AI and Trust**: As AI becomes more integral to business operations, companies that can demonstrate ethical use of AI are likely to gain a competitive edge. This involves transparent AI operations, commitment to privacy, and efforts to mitigate biases in AI algorithms. Building trust with stakeholders – customers, employees, and regulators – in the age of AI is becoming a critical aspect of competitive strategy.

GenAI is not just altering the tools businesses use but is reshaping the very grounds on which competition occurs. In this new arena, data, algorithmic prowess, adaptability, personalization, continuous innovation, and ethical practices are key battlegrounds where competitive advantages are won or lost. Understanding and mastering these new dimensions are crucial for any business seeking to thrive in the GenAI-driven landscape.

Transforming Business Strategy

The rise of GenAI is not just a technological evolution; it's a strategic revolution. In the AI era, the transformation of business strategy is profound, compelling companies to rethink and realign their approaches in several key areas:

- **Data-Driven Decision Making**: The core of modern business strategy in the GenAI era revolves around data. Companies are increasingly relying on data-driven insights generated by AI to inform their strategic decisions. This shift means that decision-making processes are no longer just based on historical data and human intuition but are augmented by predictive analytics and forward-looking insights. Organizations are investing in sophisticated data analytics tools powered by GenAI to identify market trends, understand consumer behavior, and anticipate future challenges and opportunities.

- **Innovation as a Constant**: In an environment where GenAI continuously generates new ideas and solutions, innovation becomes a persistent strategic focus. Companies are leveraging GenAI not only to enhance existing products and services but also to create entirely new offerings. This constant drive for innovation requires a shift in company culture and structure, encouraging creativity, experimentation, and agility. Businesses are forming dedicated innovation labs and teams focused on exploring AI-driven innovations, ensuring they remain at the forefront of their industries.

- **Collaboration and Ecosystems**: The complexity and sophistication of GenAI technologies necessitate new forms of collaboration and partnership. Companies are increasingly looking beyond their traditional industry boundaries to form strategic alliances with tech firms, startups, academic institutions, and even competitors. These collaborations offer access to new technologies, talent, and insights, enabling companies to leverage GenAI more effectively. In some cases, these partnerships are leading to the creation of entirely new ecosystems where data, insights, and value are shared for mutual benefit.

- **Investment in AI Talent and Infrastructure**: Recognizing the central role of GenAI in future success, companies are making significant investments in AI talent and technological infrastructure. This includes hiring AI specialists, data scientists, and ethicists, as well as investing in AI platforms and tools. Additionally, businesses are focusing on upskilling their existing workforce to be AI-savvy, ensuring that their teams can work effectively with AI technologies and concepts.

- **Ethical AI and Governance**: As AI becomes more embedded in business processes, the importance of ethical AI practices and governance increases. Companies are establishing guidelines and frameworks to ensure that their AI applications are transparent, fair, and privacy-conscious. This focus on ethical AI is not just a regulatory compliance issue but is becoming a key component of brand reputation and customer trust.

- **Agile and Adaptive Structures**: The dynamic nature of GenAI-driven markets requires businesses to adopt more agile and adaptive organizational structures. This means moving away from rigid hierarchies to more fluid and cross-functional teams that can quickly respond to new AI-driven insights and market changes. Companies are also adopting more iterative and flexible approaches to strategy development, allowing them to pivot or scale strategies in response to AI-generated insights.

The transformation of business strategy in the GenAI era is comprehensive. It requires a holistic approach that encompasses data-centric decision-making, continuous innovation, strategic collaboration, investment in talent and infrastructure, ethical AI practices, and agile organizational structures. Companies that successfully integrate these elements into their strategic planning are poised to lead in this new era of AI-driven business.

Revolutionizing Business Models

The advent of Generative AI (GenAI) is not just altering existing business processes; it's fundamentally revolutionizing the very models upon which businesses are built and operate. This transformation is evident across various aspects:

- **Personalization and Customization at Scale**: One of the most significant impacts of GenAI is its ability to offer unparalleled levels of personalization and customization in products and services. Businesses are now able to cater to individual customer preferences at a scale previously unimaginable. For instance, online retailers use GenAI to recommend products uniquely suited to each customer's buying history and preferences, while media companies provide personalized content feeds. This shift to a customer-centric model creates deeper engagement, loyalty, and value for both the business and the customer.

- **New Revenue Streams Through AI Services**: GenAI opens up novel avenues for revenue generation. Companies are now creating new business lines that offer AI-driven services such as data analytics, predictive modeling, and automated content generation. For instance, a company with expertise in AI-driven market analysis might offer these services to other businesses, creating a new source of revenue beyond its traditional product offerings.

- **Operational Efficiency and Automation**: GenAI significantly enhances operational efficiency, especially in areas like supply chain management, inventory control, and logistics. By automating routine tasks and optimizing complex processes, companies are able to reduce costs, improve speed, and increase reliability. This efficiency not only cuts down operational expenses but also enables businesses to offer more competitive pricing and faster service to their customers.

- **Dynamic and Flexible Pricing Models**: The ability to analyze vast amounts of data in real-time allows businesses to adopt dynamic and flexible pricing models. For example, airlines and hotels are using GenAI to adjust prices based on changing demand patterns, competitor pricing, and other market factors. This level of pricing flexibility enables businesses to maximize revenue and remain competitive in fluctuating market conditions.

- **Value Creation Through Data**: In the GenAI era, data itself becomes a valuable asset. Businesses are not only using data to improve their operations and offerings but also monetizing their data assets through partnerships and collaborations. For example, a company with extensive consumer behavior data might partner with a marketing firm to provide insights, thereby creating a new value stream.

- **Enhanced Customer Experiences with AI Interactions**: GenAI is enabling businesses to offer enhanced customer experiences through AI-powered interactions. Chatbots and virtual assistants, powered by sophisticated AI algorithms, provide customers with personalized, efficient service. These AI interactions can range from answering customer queries to providing shopping recommendations and technical support.

- **Sustainable and Responsible Business Practices**: GenAI also allows for more sustainable and responsible business practices. By optimizing resource use, reducing waste, and improving energy efficiency, businesses can not only reduce their environmental footprint but also meet the growing consumer demand for sustainability. This approach is becoming an integral part of business models, as consumers increasingly favor companies with strong environmental and social governance (ESG) records.

The revolution in business models brought about by GenAI is multifaceted, impacting everything from customer engagement and revenue generation to operational efficiency and sustainability. Businesses that embrace and adapt to these new models are likely to thrive in the GenAI-dominated landscape, while those that fail to evolve risk falling behind in an increasingly competitive and AI-driven world.

Optimizing Business Operations

The integration of Generative AI (GenAI) into operational processes signifies a major leap towards unprecedented efficiency and effectiveness in business operations. This transformation touches various facets of operational excellence:

- **Predictive Analytics for Inventory Management**: GenAI is revolutionizing inventory management by enabling predictive analytics. Businesses can now accurately forecast demand, optimize stock levels, and reduce waste. This capability is crucial for industries where inventory management directly impacts profitability, such as retail, manufacturing, and logistics. By predicting future trends and consumer demands, companies can ensure they have the right products in the right quantities at the right time, minimizing overstock and stockouts.

- **Optimization of Supply Chain and Logistics**: GenAI algorithms are employed to optimize supply chains and logistics operations. These AI systems can analyze complex datasets encompassing supplier performance, transportation costs, weather patterns, and market demand to identify the most efficient routes and methods for product distribution. This optimization leads to reduced lead times, lower transportation costs, and improved customer satisfaction. In logistics, for instance, GenAI can dynamically reroute shipments in real-time based on traffic conditions, weather, and other variables, significantly enhancing efficiency.

- **Enhancing Quality Control Processes**: Quality control is another area where GenAI is making a significant impact. In manufacturing and production, AI algorithms are used to detect defects and anomalies in products with a level of precision and speed that is unattainable by human inspectors. This not only improves the overall quality of the products but also reduces the cost associated with defects and returns.

- **Automating Routine and Administrative Tasks**: GenAI is automating routine and administrative tasks across various sectors, allowing human employees to focus on more complex and strategic activities. For example, in the HR sector, AI systems can handle tasks such as resume screening, initial candidate assessments, and even basic HR queries, enhancing efficiency and reducing the workload on HR staff.

- **Energy Efficiency and Resource Optimization**: In industries where energy consumption is a significant operational cost, such as manufacturing and data centers, GenAI is used to optimize energy usage and resource allocation. AI algorithms analyze patterns of energy consumption and identify ways to reduce waste and improve efficiency, leading to cost savings and a reduced environmental footprint.

- **Customer Service and Support Automation**: In customer service, GenAI-powered chatbots and virtual assistants are handling an increasing volume of queries and support tasks. These AI systems can provide quick, accurate responses to customer inquiries, improving response times and customer satisfaction. In more complex cases, they can efficiently route queries to human agents, optimizing the workload distribution.

- **Risk Management and Compliance**: GenAI also plays a crucial role in risk management and compliance. In sectors like finance and healthcare, AI systems can analyze vast amounts of data to identify potential risks and ensure compliance with regulatory requirements. For example, in banking,

AI algorithms are used for fraud detection, analyzing transaction patterns to identify and prevent fraudulent activities.

The impact of GenAI on operational excellence is profound and far-reaching. By enhancing efficiency, reducing costs, and improving quality and customer satisfaction, GenAI is enabling businesses to achieve new levels of operational excellence. As GenAI technologies continue to evolve, their role in driving operational improvements is set to grow even more significant, becoming an indispensable part of modern business operations.

Enhancing Marketing and Customer Experience

The integration of Generative AI (GenAI) in marketing and customer experience is revolutionizing the way businesses interact with and understand their customers. This transformation is not just about automation; it's about creating more personalized, engaging, and responsive experiences.

- **Personalizing Customer Interactions**: GenAI enables a level of personalization in customer interactions that was previously unattainable. By analyzing customer data, businesses can tailor their marketing messages, product recommendations, and services to individual preferences and behaviors. This personalization extends across various channels – from emails and social media to websites and mobile apps – ensuring a consistent and customized experience for each customer.

- **Conversational User Interfaces (UIs)**: A significant shift brought about by GenAI is the move from traditional UIs to conversational, AI-driven interfaces. Websites, apps, and online services are increasingly incorporating chatbots and virtual assistants that interact with users in a natural, conversational manner. These AI-powered interfaces are capable of understanding and responding to user queries, providing assistance, and even guiding users through complex processes. This shift to a conversation-based interaction model makes user experiences more intuitive, engaging, and efficient.

- **Predicting Consumer Behavior**: GenAI tools are capable of predicting consumer behavior with high accuracy. By analyzing past interactions, purchase history, browsing patterns, and other data points, AI models can anticipate future needs and preferences. This predictive capability enables businesses to proactively offer relevant products and services, enhancing customer satisfaction and loyalty.

- **Real-Time Engagement and Feedback**: GenAI facilitates real-time engagement with customers. AI algorithms can analyze customer feedback, social media interactions, and online behavior as it happens, allowing businesses to respond promptly to customer needs, trends, and sentiments. This real-time interaction is crucial for maintaining a positive brand image and for addressing any issues before they escalate.

- **Automating and Optimizing Marketing Campaigns**: GenAI is used to automate and optimize marketing campaigns, from email marketing to social media ads. AI algorithms can test different marketing messages, analyze their performance, and automatically adjust campaigns for maximum effectiveness. This automation not only saves time and resources but also ensures that marketing efforts are more targeted and successful.

- **Enhancing Customer Service**: GenAI is transforming customer service by enabling 24/7 support through AI-driven chatbots and virtual assistants. These AI tools can handle a wide range of customer service tasks, from answering FAQs to resolving basic issues, providing a seamless support experience. For more complex issues, they can intelligently route queries to the appropriate human agent, ensuring efficient resolution.

- **Creating Immersive and Interactive Experiences**: Finally, GenAI is being used to create immersive and interactive customer experiences. For example, in the retail sector, AI-powered virtual try-on tools and interactive 3D models provide customers with a richer understanding of products. In the entertainment sector, AI is used to create personalized content streams and interactive media experiences.

GenAI is fundamentally transforming marketing and customer experience, shifting the focus towards personalized, conversational, and predictive interactions. By embracing these AI-driven approaches, businesses can significantly enhance customer engagement, satisfaction, and loyalty.

Addressing GenAI Challenges

The integration of Generative AI (GenAI) into business practices, while transformative, is not without its challenges. These hurdles, however, should not deter organizations from adopting GenAI technologies, as the benefits significantly outweigh the risks. Moreover, hesitancy or delay in embracing GenAI can lead to a substantial loss of competitive edge, making it a critical strategic concern.

- **Navigating Ethical and Regulatory Complexities**: One of the primary challenges in adopting GenAI is navigating the ethical and regulatory landscape. Issues like data privacy, algorithmic bias, and ethical AI usage are at the forefront. Businesses must establish robust frameworks to ensure that their AI applications comply with legal standards and ethical norms. Despite these complexities, the benefits of GenAI in enhancing operational efficiency, driving innovation, and improving customer experiences are substantial. Proper management of these ethical and regulatory concerns can also enhance brand trust and customer loyalty.

- **Overcoming Technical and Operational Hurdles**: The technical integration of GenAI systems can be a daunting task, especially for businesses lacking in-house AI expertise. There is also the challenge of aligning AI initiatives with existing business processes and workflows. However, the efficiency gains, cost savings, and market insights offered by GenAI systems make overcoming these hurdles a worthwhile investment. Companies can seek partnerships with AI firms, invest in employee training, or hire skilled AI professionals to bridge this gap.

- **Addressing Data Security and Privacy Concerns**: With GenAI heavily reliant on data, ensuring data security and privacy becomes a significant challenge. Businesses must implement stringent data protection measures to safeguard customer information. While this requires investment in security infrastructure and governance, the data-driven insights and operational improvements gained from GenAI far outweigh these costs.

- **Keeping Pace with Rapid Technological Changes**: The fast-evolving nature of AI technology presents another challenge. Staying abreast of the latest developments and continuously updating AI systems can be resource-intensive. Nonetheless, this agility is essential in gaining and maintaining a competitive advantage in today's fast-paced market. Companies that keep pace with AI advancements can leverage these technologies to outperform competitors and enter new markets.

- **Cultural and Organizational Adaptation**: Integrating GenAI necessitates a cultural shift within the organization, fostering an environment that embraces innovation and change. This shift can be challenging but is essential for maximizing the benefits of AI. A culture that supports AI initiatives will lead to more effective implementation and utilization of these technologies.

While the challenges associated with GenAI adoption are non-trivial, the benefits significantly outweigh these difficulties. The risks of delaying or avoiding GenAI integration are substantial. Companies that fail to adopt these technologies risk losing their competitive edge, not only to existing competitors but also to new, more agile market entrants who are quick to leverage AI advantages. Thus, proactively addressing these challenges and strategically integrating GenAI is not just advisable but imperative for businesses aiming to thrive in the modern market landscape.

Case Study: AI-Powered Next-Gen Business Model

In this case study, we examine "AI-DNA Solutions", a hypothetical company that exemplifies an AI-powered next-gen business model. AI-DNA, initially a traditional software development firm, transformed into a market leader by fully integrating Generative AI (GenAI) into its operations, products, and services.

Transformation Journey

- **Initial Phase**: AI-DNA began by incorporating GenAI in its internal operations, automating routine tasks, and enhancing data analysis processes. This initial integration led to significant efficiency improvements and cost savings.

- **Expansion to Core Offerings**: The company then expanded GenAI use to its core services, developing AI-driven software solutions for clients. This included personalized software interfaces and predictive analytics tools tailored to specific industry needs.

Innovative Business Strategies

- **Data-Driven Product Development**: Leveraging GenAI, AI-DNA adopted a data-driven approach for product development, using customer feedback and market trends to guide innovation.

- **Dynamic Pricing Models**: Utilizing AI for dynamic pricing, AI-DNA could adjust its service charges in real-time based on demand, competition, and customer profile.

GenAI in Customer Engagement and Marketing

- **Personalized Marketing**: AI-DNA used GenAI to analyze customer data, creating highly personalized marketing campaigns that significantly increased engagement and conversion rates.

- **Enhanced Customer Service**: Implementing AI-powered chatbots, AI-DNA offered 24/7 customer service, handling inquiries efficiently and escalating complex issues to human representatives.

Operational Excellence

- **Supply Chain Optimization**: AI-DNA used GenAI for supply chain management, predicting demand for services and optimizing resource allocation.

- **Human Resource Management**: AI-driven tools were employed for talent acquisition and management, enhancing the efficiency and effectiveness of the HR department.

Overcoming Challenges

- **Navigating Ethical Concerns**: AI-DNA established a strict ethical framework for AI use, ensuring transparency and fairness in customer interactions and data handling.

- **Continuous Learning and Adaptation**: The company invested in ongoing training for staff to keep up with AI advancements, fostering a culture of continuous learning and innovation.

Impact and Results

- **Market Position**: By fully integrating GenAI, AI-DNA Solutions positioned itself as a leader in innovative software solutions, outpacing traditional competitors.

- **Financial Growth**: The company saw a significant increase in revenue, driven by improved operational efficiency, customer retention, and the introduction of new AI-driven services.

- **Customer Satisfaction**: Personalization and improved service quality led to higher customer satisfaction and brand loyalty.

AI-DNA Solutions' journey illustrates the transformative impact of GenAI on a traditional business. By embracing AI-driven innovations in operations, customer engagement, and product development, TechNovate not only enhanced its efficiency and market position but also set new industry standards. This case study serves as a blueprint for other businesses looking to leverage AI for competitive advantage and underscores the importance of strategic integration of GenAI in all business facets.

Analyzing GenAI Business Cases Across Industries

The application of Generative AI transcends industry boundaries, offering transformative solutions tailored to the unique challenges and opportunities of each sector. This section delves into various industries, illustrating how GenAI is not only enhancing existing processes but also enabling new strategies and business models.

Healthcare: Revolutionizing Patient Care and Research

- **Personalized Medicine**: GenAI is facilitating a shift towards personalized medicine, where treatments are tailored to individual patient profiles, greatly improving outcomes.

- **Drug Discovery and Development**: Accelerating the pace of drug discovery, GenAI can analyze vast datasets to identify potential drug candidates and predict their effectiveness, reducing the time and cost of bringing new drugs to market.

- **Diagnostic Accuracy**: In diagnostics, GenAI enhances the ability to interpret medical imaging, leading to earlier and more accurate disease detection.

Finance: Enhancing Risk Management and Customer Services

- **Algorithmic Trading**: GenAI drives sophisticated trading strategies, analyzing market data for predictive insights, leading to more informed trading decisions.

- **Risk Assessment and Fraud Detection**: By analyzing transaction patterns, GenAI improves the detection of fraudulent activities and enhances credit risk assessment.

- **Personalized Financial Advice**: Through AI-powered robo-advisors, GenAI is democratizing access to personalized financial advice, tailoring investment strategies to individual investor profiles.

Retail: Transforming Customer Experience and Operations

- **Customer Behavior Analysis and Personalization**: Retailers are using GenAI to analyze customer behavior, enabling personalized marketing and product recommendations.

- **Inventory Management**: GenAI aids in predictive inventory management, ensuring optimal stock levels are maintained, reducing waste and lost sales.

- **Supply Chain Optimization**: Enhancing supply chain efficiency through predictive logistics and demand forecasting.

Manufacturing: Driving Efficiency and Innovation

- **Predictive Maintenance**: GenAI predicts equipment failures, reducing downtime and maintenance costs.

- **Production Optimization**: It optimizes production processes, enhancing efficiency and reducing waste.

- **Quality Control**: GenAI improves quality control measures, automating inspections and detecting defects more accurately than human operators.

Education: Personalizing Learning Experiences

- **Customized Learning Paths**: GenAI facilitates personalized learning, adapting educational content to individual student needs and learning styles.

- **Automated Grading and Feedback**: AI tools assist educators by grading assignments and providing feedback, enhancing the learning process.

Automotive: Fueling Advances in Safety and Efficiency

- **Autonomous Vehicles**: GenAI is central to the development of self-driving technologies, promising to revolutionize transportation safety and efficiency.

- **Supply Chain and Manufacturing Optimization**: Enhancing supply chain logistics and manufacturing processes through predictive analytics and automation.

Entertainment and Media: Curating Personalized Experiences

- **Content Recommendation**: In media and entertainment, GenAI curates personalized content, enhancing viewer engagement.

- **Interactive and Immersive Experiences**: It is creating new forms of interactive media, offering unique and immersive experiences to users.

Agriculture: Optimizing Production and Sustainability

- **Crop Management and Yield Prediction**: GenAI helps in predicting crop yields, optimizing irrigation and pest control, leading to increased efficiency and sustainability in farming practices.

Energy: Streamlining Consumption and Distribution

- **Energy Consumption Optimization**: GenAI is used in optimizing energy consumption patterns in industrial and residential settings, contributing to sustainability efforts.

- **Predictive Maintenance in Energy Infrastructure**: Enhancing the reliability and efficiency of energy production and distribution through predictive maintenance.

The diverse applications of GenAI across these industries underscore its role as a pivotal driver of innovation and efficiency. The business cases in each sector reveal GenAI's potential to not only improve existing processes but also to redefine industry paradigms, creating new opportunities for growth and development. For businesses in these industries, the adoption of GenAI is rapidly moving from a competitive advantage to a necessity for survival and success in an increasingly AI-driven world.

GenAI Across Industries: A Unified Approach

The integration of Generative AI (GenAI) is not confined to specific sectors; rather, it presents a unifying technological advancement with applications spanning across various industries. This section explores how the principles and applications of GenAI create a bridge across diverse sectors, fostering innovation, efficiency, and growth in a multitude of fields.

Cross-Industry Synergies and Shared Learnings

- **Transfer of Knowledge and Best Practices**: Innovations and advancements in GenAI in one industry can inform and inspire applications in another. For instance, GenAI techniques used for predictive maintenance in manufacturing can be adapted for predictive analytics in healthcare or finance.

- **Collaborative Ecosystems**: The development of GenAI fosters collaborative ecosystems, where companies, researchers, and technologists from different industries share insights, data, and strategies. This cross-pollination of ideas accelerates the advancement and refinement of GenAI applications.

Scalability and Adaptability of GenAI Solutions

- **Scalable Solutions Across Sectors**: GenAI offers scalable solutions that can be adapted to different scales and scopes, making it relevant for businesses of all sizes and across sectors. For example, a small retail business and a large healthcare provider can both use GenAI-driven data analysis to enhance their operations, albeit in different contexts.

- **Adapting GenAI for Sector-Specific Needs**: While GenAI has universal applications, its real power lies in its adaptability to meet specific industry needs. Customization of GenAI algorithms allows for tailored solutions, whether it's for optimizing energy use in industrial settings or enhancing customer engagement in the service sector.

Unifying Frameworks and Standards

- **Development of Universal Standards**: As GenAI becomes ubiquitous across industries, there is a growing need for universal standards and frameworks to ensure compatibility, interoperability, and ethical use of AI technologies.

- **Regulatory Harmonization**: A unified approach to GenAI also involves aligning regulatory landscapes across different industries, ensuring that the deployment of AI technologies complies with global standards and ethical considerations.

Leveraging GenAI for Broader Societal Impact

- **Addressing Global Challenges**: GenAI has the potential to address broad societal challenges, such as climate change, healthcare, and education. By bridging industry divides, GenAI can be leveraged for large-scale impact, offering solutions that transcend commercial benefits and contribute to societal well-being.

- **Enhancing Global Connectivity and Collaboration**: GenAI serves as a catalyst for global connectivity, fostering collaborations that transcend geographical and industrial boundaries. This global network enhances the exchange of knowledge, resources, and technologies, driving collective progress.

In essence, Bridging GenAI Across Industries highlights the interconnectedness of this transformative technology across different sectors. It emphasizes the importance of shared learnings, collaborative efforts, and the development of universal standards to maximize the benefits of GenAI. By adopting a unified approach to GenAI, industries can not only enhance their individual operations and services but also contribute to broader societal and global advancements. This cross-industry synergy ensures that the benefits of GenAI are widely distributed, driving innovation and growth on a global scale.

Strategic Imperative: Confronting the GenAI Adoption Urgency

In the dynamic realm of business technology, the decision to adopt Generative AI (GenAI) is not just about embracing a new tool; it's about making a strategic leap into the future. We underscores the critical nature of immediate action in integrating GenAI, highlighting the substantial risks of delay and the compelling need to embrace these technologies now.

The Risk of Delay

Delaying the adoption of GenAI carries significant and multifaceted risks. In an environment where technological advancements occur at an unprecedented pace, a cautious wait-and-see approach can be detrimental:

- **Falling Behind Competitors**: As GenAI redefines the boundaries of innovation and efficiency, competitors who harness its capabilities will move ahead. They will not only innovate faster but also achieve greater operational efficiencies, and deliver superior customer experiences. This advantage is not just in terms of product offerings but also in the agility to respond to market changes and customer needs.

- **Market Disruption by New Entrants**: The risk extends beyond current competitors. The market today is rife for disruption by new, AI-savvy players who can leverage GenAI to create novel business models and strategies. These entities, unencumbered by legacy systems and traditional approaches, can rapidly capture market share, particularly from incumbents slow to adapt to the GenAI revolution.

Embracing the Future Today

The integration of GenAI represents more than keeping pace with technological trends; it signifies a fundamental shift in business paradigms:

- **Preparing for Future Advancements**: Adopting GenAI is a proactive step towards future-proofing a business. It's not merely about the immediate gains but about preparing for the next wave of AI advancements and other emerging technologies. Companies that start building their GenAI capabilities now will be better positioned to leverage future technologies, maintaining a continuous edge in innovation.

- **Building a Foundation for Continuous Adaptation and Growth**: Embracing GenAI is about laying the groundwork for an adaptable, resilient, and growth-oriented business model. It involves cultivating a culture that is responsive to technological advancements, fostering a workforce skilled in new technologies, and establishing processes that are flexible and scalable. This foundation is crucial for thriving in an AI-driven business landscape, where change is the only constant.

- **Transforming Business Operations and Competitiveness**: The immediate integration of GenAI transforms not just specific operations but the entire competitive landscape of a business. It enables companies to redefine their market strategies, operational efficiencies, customer engagement models, and even their core business models. This transformation is comprehensive, affecting every aspect of how a business operates and competes.

In conclusion, the imperative of immediate action in adopting GenAI cannot be overstated. The risks of delay are clear and present, with the potential for significant competitive disadvantage and market disruption. On the other hand, embracing GenAI today prepares businesses for not just the current landscape but also for future advancements and challenges. It is a strategic decision that goes beyond mere technology adoption, representing a commitment to continuous innovation, adaptation, and growth in an increasingly AI-centric world.

Chapter 2

Conventional AI to AI-First to AI-Native: A Paradigm Shift

"A paradigm shift occurs when old ways of thinking are no longer effective or useful, and new ways of thinking are necessary for progress." — Ken Robinson

The integration of artificial intelligence (AI) into business operations is undergoing a significant paradigm shift. Companies are transitioning from an AI-First approach, where AI is added to enhance processes, to an AI-Native model where AI forms the core business infrastructure. This evolution promises to redefine enterprise capabilities and competitive advantage.

This chapter explores the differences between conventional AI approaches, AI-First strategies, and the emerging AI-Native paradigm. We analyze the capabilities unlocked at each level and the infrastructure needed to facilitate this digital transformation.

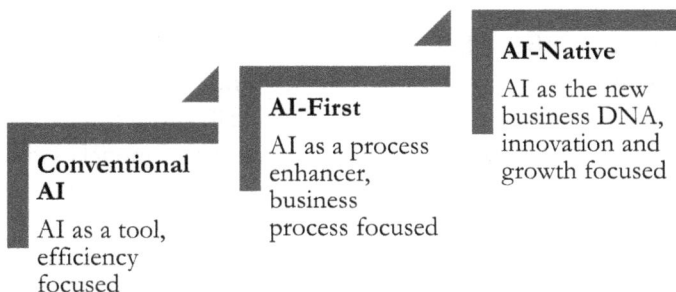

AI-Native
AI as the new business DNA, innovation and growth focused

AI-First
AI as a process enhancer, business process focused

Conventional AI
AI as a tool, efficiency focused

Figure 2-1: Three Distinct AI Paradigms in Business

Conventional AI: A Technology Perspective

The traditional view of AI in business treats it as a standalone technology, primarily used for automating routine tasks such as customer service via chatbots, process automation, and data entry. While these implementations offer short-term benefits, they significantly limit AI's transformative potential:

- **Narrow Applications**: Conventional AI is often pigeonholed into single-task applications, ignoring broader, enterprise-wide integration. This limits exploration into more impactful, innovative uses. For instance, while a chatbot might handle customer queries, it's not leveraged for deeper analytics or customer insights.

- **Peripheral Role**: In this approach, AI functions more as an add-on rather than a core business component. It operates in isolation, detached from other systems and data sources, which restricts its capabilities. An example is using AI solely for data entry without integrating it with analytics tools for strategic decision-making.

- **No Fundamental Change**: AI under this paradigm digitizes certain processes but doesn't alter the overall business model. Companies modernize specific tasks but cling to legacy structures. For instance, AI might be used to streamline inventory management, but the overall supply chain system remains unchanged.

- **Short-Term Cost Savings**: The primary driver here is operational cost savings through the automation of manual tasks. This focus misses out on long-term opportunities for evolving business models. An example is automating customer service responses without leveraging AI for customer relationship building or retention strategies.

- **Lack of Feedback Loops**: Traditional AI implementations often lack mechanisms for continuous learning and improvement. AI models, trained on limited datasets in isolation, fail to utilize more diverse data sources for enhanced learning. For example, an AI model used for forecasting sales might not adapt to changing market trends due to lack of dynamic data inputs.

While the conventional AI approach enhances short-term efficiencies, it falls short in harnessing AI for long-term innovation and growth. AI remains a peripheral tool rather than an integral part of the company's DNA. Business processes might be digitized on the surface, but the overall strategy still adheres to outdated, AI-un-suitable mindsets. This outdated perspective limits the scope of possibilities, rather than expanding them.

AI-First: Strategic Business Enhancement Approach

Definition: AI-First refers to a strategic approach where companies recognize AI's vast potential and take significant steps to integrate AI across various business units to enhance existing processes, products, and services.

The AI-First strategy represents a significant shift in how businesses view and utilize artificial intelligence. This approach is grounded in the recognition of AI's extensive capabilities, prompting organizations to integrate it across various business units, thereby enhancing processes, products, and services.

In the AI-First paradigm, businesses begin to see AI not just as a tool, but as a powerful component that can be integrated into their existing infrastructure. The analogy can be drawn to adding a high-performance engine to an existing car – the upgrade makes the car perform better, but the car's fundamental design remains unchanged. In adopting an AI-First strategy, companies use AI to boost efficiency, automate routine operations, and enrich customer interactions. Yet, AI is still one part of a larger technological ecosystem.

Key Characteristics

1. **Enhanced Integration of AI**: In an AI-First strategy, AI is not just an isolated tool but is integrated across different business units. This integration is designed to enhance existing processes, products, and services. The approach recognizes AI's ability to add value in various aspects of the business, from customer interaction to back-end operations.

2. **Operational Improvement Focus**: The primary goal is to use AI to improve operational efficiency. This includes automating routine and repetitive tasks, thus freeing up human resources for more complex and strategic activities. The emphasis is on making existing processes faster, more accurate, and more cost-effective.

3. **AI as Part of a Broader Technological Toolkit**: While AI plays a significant role in the AI-First approach, it is considered one of many tools in the technological arsenal of a company. It complements other technologies

and strategies, ensuring a balanced and integrated approach to innovation and problem-solving.

4. **Enhancing Existing Infrastructure**: The AI-First approach does not necessitate a complete overhaul of existing systems. Instead, AI is used to improve what is already in place, akin to upgrading a car with a more powerful engine. The existing business infrastructure is thus optimized rather than replaced.

5. **Process Enhancement Without Fundamental Business Model Change**: In this strategy, AI leads to significant enhancements in processes, but the core business model of the organization generally remains unchanged. The focus is on improving how things are done rather than rethinking what is being done.

6. **AI as an Accelerator, Not a Core Driver**: AI in an AI-First approach is seen more as an accelerator of business functions rather than the primary driver. It boosts the performance of various operations but within the framework of the existing business model and strategy. There is recognition of AI's potential, but it has not yet reshaped the entire organizational strategy or structure.

Examples in Practice

- **Retail Sector**: A retail company might employ AI for customer service through advanced chatbots, which not only respond to queries but also provide personalized shopping recommendations based on customer behavior.

- **Finance Industry**: Financial institutions might use AI for more sophisticated fraud detection systems, where AI algorithms continuously learn and adapt to new fraudulent patterns, significantly reducing the risk of financial crimes.

- **Omnichannel Customer Engagement**: Companies are employing AI to create seamless customer experiences across various channels like web, mobile, and physical stores. AI-powered tools offer personalized recommendations and predictive analytics to enhance customer engagement.

- **Operational Efficiency**: AI is being used to streamline operations in areas like supply chain management, logistics, and manufacturing. By automating repetitive tasks and employing predictive models, businesses can optimize resource allocation and reduce downtime.

- **Strategic Business Insights**: Through tools like fraud analytics, sentiment analysis, and quantitative modeling, AI provides businesses with deeper insights into market trends, customer preferences, and workflow optimization.

- **Content Personalization**: In media and entertainment, AI is used to curate personalized content recommendations, enhancing user engagement and loyalty.

While the AI-First approach marks a considerable advancement in the application of AI, it typically does not alter the core business model. AI acts as a powerful accelerator, enhancing various operations, but the fundamental structure and strategy of the business often remain intact. The technology, though increasingly integrated, is not yet fully woven into the organization's fabric. This approach represents a transitional phase, where AI is recognized for its potential and begins to play a more significant role, yet the full transformative integration of AI into the business model is still on the horizon.

AI-Native: A Paradigm for Business Transformation

> **Definition:** AI-Native embodies a mindset and paradigm where artificial intelligence (AI) principles are deeply integrated into the fabric of decision-making, problem-solving, and creative processes. It transcends traditional approaches, prioritizing AI-centric methods in both personal and professional realms, and fostering a culture that views AI as a fundamental component of intellectual and practical endeavors.

AI-Native signifies a profound transformation in the business landscape, where AI principles are not just integrated but deeply embedded into the very core of decision-making, problem-solving, and creative processes. This approach transcends traditional methods, placing AI at the heart of both personal and professional realms and fostering a culture that sees AI as an indispensable element of business.

The evolution from AI-First to AI-Native is not merely an upgrade; it's a fundamental change in the business ethos. AI-Native is the point where AI ceases to be just a part of business operations and becomes the central element around which businesses are built and run. This shift necessitates a complete reevaluation of strategies, business models, and operations, leading to a future where businesses are agile, innovative, and fully equipped to utilize AI's capabilities for market leadership.

AI-Native Ecosystem

The AI-Native ecosystem represents a comprehensive and holistic integration of AI into every facet of a business, fundamentally altering how organizations operate, innovate, and compete. This ecosystem is characterized by its depth of integration, agility, and the transformative impact it has on both the technological and human elements of the enterprise.

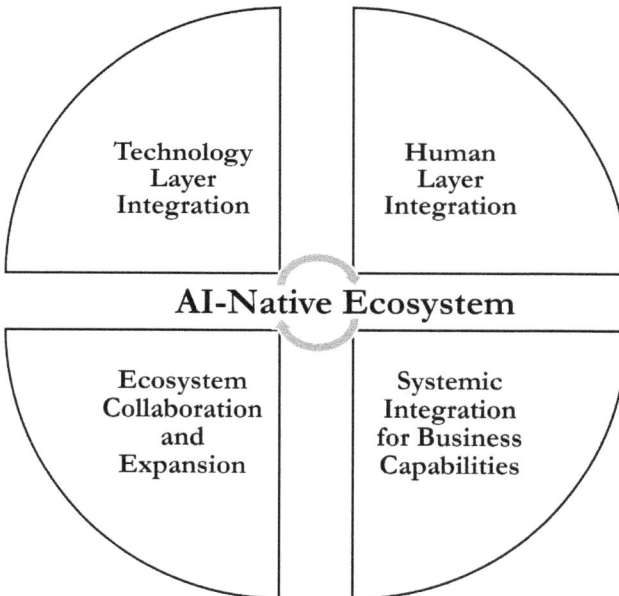

Technology Layer Integration

Human Layer Integration

AI-Native Ecosystem

Ecosystem Collaboration and Expansion

Systemic Integration for Business Capabilities

Figure 2-2: The Four Core Dimensions of AI-Native Ecosystem

1. Technology Layer Integration

- **Unified AI Infrastructure**: In an AI-Native ecosystem, AI is the backbone of all technological infrastructure. This involves a seamless connection of databases, applications, APIs, cloud services, and IoT devices. AI algorithms operate across this integrated network, allowing for real-time data processing and decision-making.

- **Data-Driven Operations**: The technology stack in an AI-Native company is designed to be data-centric. AI models continuously analyze data streams from various sources, offering insights that drive operational efficiency, product development, and customer engagement strategies.

- **Scalable and Flexible Architecture**: The technological infrastructure is scalable, able to handle increased data loads, and flexible to adapt to new AI advancements. This agility is crucial for businesses to stay competitive in rapidly changing markets.

2. Human Layer Integration

- **AI-Enabled Workforce**: In the AI-Native ecosystem, the workforce is not only familiar with AI tools but also skilled in leveraging these tools for enhanced productivity and creativity. Training programs and workshops are regular features, ensuring that employees are up-to-date with the latest AI technologies and methodologies.

- **Decision-Making and Problem-Solving**: AI tools assist in decision-making processes, offering predictive insights and scenario analysis. This integration leads to more informed and strategic decisions across all levels of the organization.

- **Innovation Culture**: An AI-Native company fosters a culture of innovation where employees are encouraged to experiment with AI applications. This generative culture supports a continuous cycle of ideation, prototyping, and implementation.

3. Systemic Integration for Business Capabilities

- **End-to-End Process Optimization**: AI is integrated across all business processes, from supply chain management to customer service, ensuring that each process is optimized for efficiency and effectiveness.

- **Real-Time Analytics and Reporting**: The AI-Native ecosystem provides real-time analytics and reporting capabilities, enabling businesses to respond quickly to market trends, customer behavior, and operational challenges.

- **Customization and Personalization**: AI algorithms analyze customer data to provide personalized experiences and products, enhancing customer satisfaction and loyalty.

- **Predictive and Prescriptive Insights**: AI not only forecasts future trends and behaviors but also prescribes actionable strategies to capitalize on these insights, keeping the business ahead of the curve.

4. Ecosystem Collaboration and Expansion

- **Interconnected Business Networks**: AI-Native businesses often form interconnected networks with partners, suppliers, and customers, facilitated by AI-driven APIs and platforms. This collaboration leads to a more robust and resilient business ecosystem.

- **Market Adaptability and Responsiveness**: The AI-Native ecosystem is inherently adaptable, allowing businesses to quickly respond to changes in the market, regulatory environments, and customer preferences.

The AI-Native ecosystem is a dynamic, interconnected, and continuously evolving environment where AI is not just a component but the driving force. This ecosystem redefines the relationship between technology and business, leading to unparalleled levels of efficiency, innovation, and market responsiveness.

AI-Native Enterprise

In the rapidly evolving landscape of artificial intelligence (AI), AI-Native enterprises stand at the forefront, pioneering a future where AI is not just an enabler but the core essence of business operations. These organizations represent the zenith of AI integration in business, redefining traditional models and strategies through the transformative power of AI. They are characterized by their deep-rooted commitment to leveraging AI across all aspects of their operations, from strategic decision-making to day-to-day tasks, setting new benchmarks in innovation, efficiency, and customer engagement. This section delves into the core characteristics of AI-Native enterprises, illustrating how they are reshaping the business world.

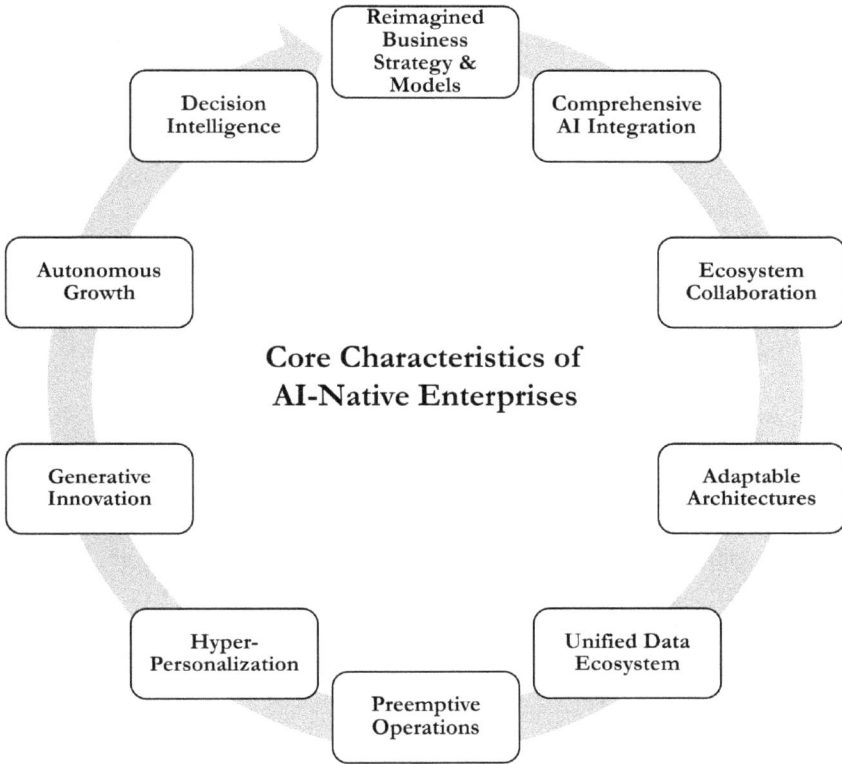

Figure 2-3: Top 10 Characteristics of AI Native Enterprises

Core Characteristics of AI-Native Enterprises

1. **Reimagined Business Strategy and Models:** AI-Native companies have transformed the role of AI from an auxiliary tool to a central business driver, deeply ingrained in their organizational DNA. This paradigm shift involves a fundamental realignment of business strategies and models around AI, aiming to unlock exponential value. For instance, a company might evolve from a traditional product-based model to a service-oriented approach, utilizing AI to offer highly personalized services. This strategic evolution leads to the creation of platform ecosystems, where data, services, and products are seamlessly interwoven, providing comprehensive and integrated solutions to customers.

2. **Comprehensive AI Integration:** In an AI-Native environment, AI's integration transcends mere widespread application; it becomes the foundation of all business processes. From product development to supply chain management, AI-driven insights are central to decision-making. This

integration enables a comprehensive view of business operations, where every function and decision is influenced and enhanced by AI.

3. **Ecosystem Collaboration:** Collaboration in AI-Native companies breaks the mold of traditional business boundaries, fostering symbiotic relationships with a broad spectrum of partners, suppliers, and even competitors. This collaboration often involves sharing data through APIs, leading to richer, more comprehensive solutions and joint AI-driven research and development initiatives. The result is a thriving innovation ecosystem that opens up new market opportunities and fosters a more resilient business model.

4. **Adaptable Architectures:** The technological infrastructure of AI-Native companies is a testament to adaptability and resilience. These architectures are not static; they are agile, scaling and evolving with the changing landscape of AI technology and market demands. This adaptability is evident in their use of cloud computing resources capable of handling extensive AI processing, data storage systems managing diverse data influx, and interconnected networks ensuring seamless data flow. This flexible infrastructure keeps the company at the technological forefront, ready to integrate new AI advancements as they emerge.

5. **Unified Data Ecosystem:** In AI-Native organizations, data is not isolated in silos but is part of a unified ecosystem. This approach allows AI systems to access and analyze comprehensive datasets, including customer information, operational data, and market trends. Such a unified data strategy enhances the accuracy and predictive power of AI models, providing a holistic view of the business and its environment. The result is more effective and informed decision-making processes, driven by high-quality, integrated data.

6. **Preemptive Operations:** AI-Native businesses excel in their ability to forecast market changes and anticipate trends. They leverage AI not just for reactive measures but for proactive strategy adaptation. Whether adjusting supply chain operations in response to consumer demand fluctuations or shifting marketing strategies to align with emerging trends, these companies use AI to stay ahead of the curve, ensuring effective resource allocation and strategic risk management.

7. **Hyper-Personalization:** The level of personalization achieved by AI-Native companies is unparalleled. By tapping into enterprise-wide data, these companies can tailor their products and services to individual customer

preferences and behaviors. This hyper-personalization extends beyond product recommendations to include personalized marketing campaigns, custom product offerings, and individualized customer journey mapping. The outcome is a significant enhancement in customer engagement and loyalty, as experiences are tailored to the unique needs and preferences of each customer.

8. **Generative Innovation:** Innovation in AI-Native companies is dynamic and ongoing. AI is not only used to refine existing products and services but also as a tool for generating new ideas and opportunities. This generative approach to innovation is embedded in the company culture, where AI is seen as a partner in the creative process. Employees are encouraged to experiment and learn, fostering a culture of rapid ideation and 'fail-fast' innovation, continually pushing the boundaries of what's possible.

9. **Autonomous Growth:** Growth in AI-Native companies is characterized by its self-driven nature. AI systems autonomously optimize processes and identify opportunities, minimizing the need for human intervention in routine tasks. This autonomy enables businesses to scale rapidly and efficiently, focusing human resources on strategic and creative tasks, thus driving growth and innovation.

10. **Decision Intelligence:** Decision-making in AI-Native companies is significantly enhanced by AI-driven intelligence. AI tools analyze vast amounts of data to provide actionable insights, informing strategic and operational decisions. This intelligence enables companies to make informed choices, predict outcomes, and identify potential risks, thereby enhancing the overall effectiveness of their strategies and operations.

AI-Native enterprises are not just adapting to the AI revolution; they are actively shaping it. By deeply embedding AI into every facet of their business, these organizations are redefining what it means to be a leader in the digital age. They demonstrate how AI can be harnessed not just as a tool for incremental improvement, but as a transformative force that reimagines business models, revolutionizes customer experiences, and drives unprecedented growth and innovation. As we look towards the future, AI-Native companies serve as beacons, illuminating the path for others to follow and showing the immense possibilities that arise when AI is fully integrated into the business fabric. Their journey represents a significant leap in the evolution of business, mirroring the impact of major technological shifts of the past and setting new standards for the future of industry and commerce in the AI era.

Strategic Advantages of AI-Native Transformation

In the realm of business leadership, embracing an AI-Native strategy signifies a profound transformation that transcends mere technological advancement. This approach redefines the strategic landscape for executive leaders and board members, offering a multifaceted suite of capabilities that drive agility, innovation, and sustainable growth. By integrating AI at the core of business operations and decision-making, AI-Native companies unlock unparalleled opportunities for strategic development, operational excellence, and market leadership. This section delves into the various ways AI-Native transformation empowers executive decision-making and business capabilities, underlining its critical role in shaping the future trajectory of organizations.

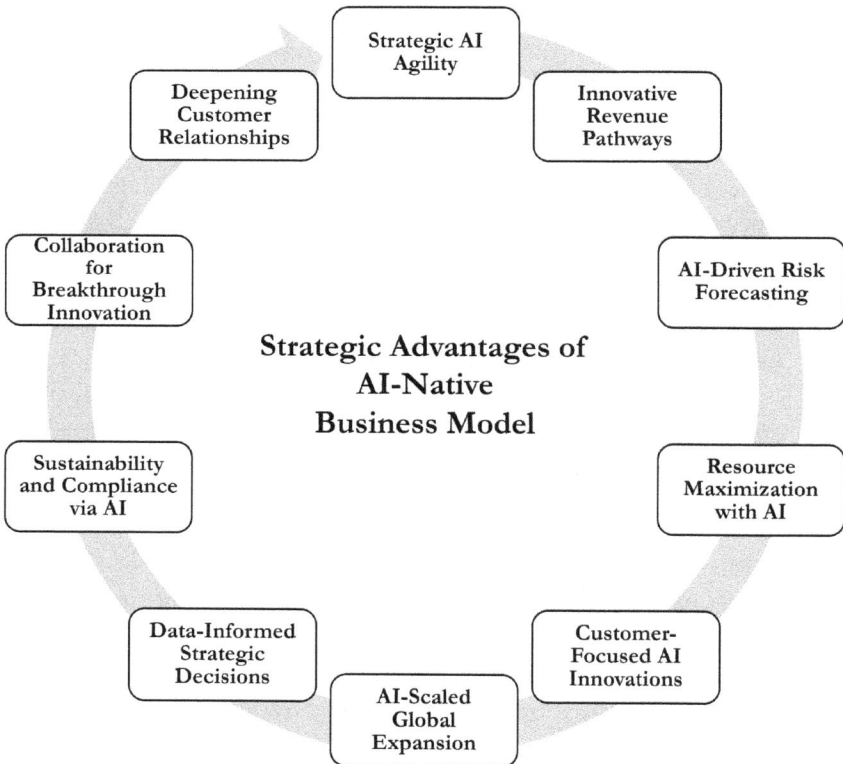

Figure 2-4: Top 10 Strategic Advantages of AI-Native Business Model

Strategic Advantages in the AI-Native Business Model

1. **Strategic AI Agility**: From the vantage point of executive leaders and board members, the AI-Native approach elevates a company's ability to nimbly navigate market trends. The advanced predictive analytics powered by AI enable swift strategic adjustments, keeping the business ahead of market curves and competitors. This agility is crucial for maintaining competitive advantage and responding effectively to fast-evolving market conditions.

2. **Innovative Revenue Pathways**: AI-Native companies stand out in their ability to identify and develop new revenue streams, a key concern for top executives. Utilizing AI to decipher complex data patterns, these companies are adept at uncovering untapped market opportunities and customer needs, leading to the creation of innovative products and services. This strategic diversification of revenue sources is vital for long-term financial stability and growth.

3. **AI-Driven Risk Forecasting**: For executive leaders, risk management is a paramount concern. The integration of AI across business operations substantially enhances this aspect by enabling predictive risk assessment. AI's ability to analyze diverse datasets for potential risks empowers proactive decision-making, crucial for mitigating losses and optimizing business strategies.

4. **Resource Maximization with AI**: AI-Native firms achieve a high level of resource optimization, a key value for executives looking to maximize ROI. AI's real-time operational data analysis ensures the most efficient use of resources, leading to significant cost savings and better allocation to high-yield areas. This efficiency is critical for driving profitability and operational excellence.

5. **Customer-Focused AI Innovations**: Executives in AI-Native companies harness AI for deep customer insights, ensuring that product development is closely aligned with market demands and customer preferences. This customer-centric approach in design and functionality is essential for increasing market share and customer loyalty.

6. **AI-Scaled Global Expansion**: For leaders aiming for global reach, AI-Native companies offer an effective model. AI-driven scalability facilitates smoother expansion into new markets, with automated processes and informed strategies ensuring consistent quality and regional compliance.

This global scalability is key for businesses seeking to broaden their market footprint.

7. **Data-Informed Strategic Decisions**: AI's role in strategic decision-making is of immense value to executive leaders. AI tools provide deep, actionable insights into market dynamics, operational performance, and future growth trajectories. This data-driven approach to strategy formulation equips executives and board members with the comprehensive understanding necessary for making well-informed, impactful decisions.

8. **Sustainability and Compliance via AI**: In today's business world, sustainability and regulatory compliance are critical for long-term success. AI assists in developing sustainable practices by optimizing resource usage and reducing waste, aligning with environmental goals. Additionally, AI's regulatory monitoring ensures compliance across different global markets, a vital aspect for maintaining corporate integrity and avoiding legal pitfalls.

9. **Collaboration for Breakthrough Innovation**: Executive leaders recognize the value of innovation in sustaining business growth. AI-Native organizations foster a culture of collaborative innovation, partnering with diverse entities for shared knowledge and creativity. This collaborative approach, facilitated by AI, leads to breakthroughs and shared success, expanding the company's innovation capacity beyond its internal resources. Such partnerships are instrumental in driving industry advancements and keeping the company at the forefront of technological progress.

10. **Deepening Customer Relationships Through AI**: Beyond the operational and strategic benefits, AI-Native companies leverage AI to build and maintain deeper customer relationships. This aspect is particularly valuable for executive leaders focusing on long-term customer engagement and brand loyalty. AI's continuous learning from customer interactions and feedback refines engagement strategies, fostering stronger, more meaningful relationships that contribute to brand loyalty and customer retention.

For executive leaders and board members, adopting an AI-Native strategy is far more than a technological upgrade; it is a strategic imperative. This approach aligns closely with the core objectives of executives, providing a robust foundation for agility, innovation, and sustainable growth. In an increasingly competitive business landscape, AI-Native companies are poised to lead, leveraging the full spectrum of AI capabilities to redefine market dynamics, enhance operational efficiency, and foster enduring customer relationships. As such, the AI-Native model stands as a beacon for forward-thinking businesses aiming to secure a prominent place in the future of industry and commerce.

Case Study: An AI-Native Organization

TransNative Solutions is a fictional, forward-thinking organization that has embraced the AI-Native model, transforming its business operations, strategy, and culture. The company operates in the technology sector, providing innovative solutions across various industries, including healthcare, retail, and finance.

The AI-Native Transformation

- **Strategic AI Integration**: TransNative Solutions underwent a strategic overhaul, placing AI at the center of its business model. This shift involved redesigning its product offerings to be AI-driven, leveraging AI to enhance customer experiences and optimize operations.

- **Innovative Business Models**: Embracing an AI-Native approach, TransNative Solutions moved from traditional product-based offerings to service-oriented solutions. For example, in healthcare, they shifted from selling medical devices to offering AI-powered diagnostic and treatment planning services.

- **Data-Centric Operations**: At TransNative, every decision is data-driven. AI algorithms analyze market trends, customer feedback, and operational data, providing insights for strategic decisions. This approach has enabled the company to stay ahead of market trends and continuously adapt its offerings.

- **Collaborative Ecosystems**: Recognizing the importance of collaboration, TransNative has formed partnerships with various entities, including academic institutions for AI research and development, and other businesses for data sharing and joint ventures.

Impact of AI-Native Model

- **Enhanced Customer Experiences**: By leveraging AI, TransNative has been able to offer personalized services to its clients. For instance, in retail, their AI systems analyze customer data to provide tailored shopping experiences and recommendations.

- **Operational Efficiency**: AI-driven automation of routine tasks has significantly increased operational efficiency at TransNative. This includes everything from automated customer service interactions to AI-managed supply chains.

- **Innovation and Market Responsiveness**: The AI-Native model has fostered a culture of innovation at TransNative. The company regularly introduces new services and products, quickly adapting to changing market demands and customer needs.

- **Global Expansion**: AI scalability has enabled TransNative to efficiently expand into new markets. Their AI systems are adept at handling regional compliance and localization, making global operations smoother and more effective.

- **Sustainable Practices**: TransNative has utilized AI in optimizing resource usage and minimizing waste, contributing to its sustainability goals. AI algorithms help in efficiently managing energy consumption and reducing the carbon footprint of their operations.

Challenges and Solutions

- **Adapting to Rapid Change**: The fast-paced evolution of AI technology posed a challenge for TransNative. The company addressed this by investing in continuous learning and development programs for its employees, ensuring they stay abreast of the latest AI advancements.

- **Data Privacy and Security**: Handling vast amounts of data, TransNative faced challenges around data privacy and security. They implemented robust AI-driven security protocols and adhered to strict data governance standards to protect sensitive information.

- **Integrating AI Across Departments**: Initially, integrating AI across various departments was challenging due to differing levels of AI readiness. TransNative addressed this by implementing cross-departmental training and establishing an internal AI advisory board to oversee the integration process.

TransNative Solutions represents an ideal example of an AI-Native company, demonstrating the transformative impact of fully integrating AI into business operations and strategy. Their journey highlights the potential benefits, including enhanced efficiency, innovation, and customer engagement, as well as the challenges and solutions inherent in adopting an AI-Native model. As a case study, TransNative provides valuable insights for other organizations aspiring to transition to an AI-Native model.

AI Spectrum in Business: A Comparative Analysis

In the dynamic business world, understanding the spectrum of AI integration—from Conventional AI through AI-First to AI-Native—is essential for companies seeking to harness the full power of artificial intelligence. This comparative analysis illuminates the evolution of AI in the business sector, showcasing how its role has transformed from a functional tool to a central pillar in business strategy and innovation.

Conventional AI: Focuses on automating routine tasks for short-term cost savings and efficiency. It lacks deep integration with business strategies and processes, leading to limited applications and innovation.

AI-First: Represents a strategic shift where AI is used to enhance and optimize business processes and customer experiences. While more integrated than conventional AI, it does not fundamentally alter the business model or structure.

AI-Native: Embodies a transformational shift where AI is deeply integrated into every aspect of the business, from decision-making to operations. It fosters a culture of innovation, leads to the development of new business models, and leverages AI for transformative growth.

Aspect	Conventional AI	AI-First	AI-Native
Core Philosophy	AI as a tool for automation of routine tasks	AI as an enhancer of existing processes	AI as a fundamental, integrated business driver
Application Scope	Narrow, task-specific applications	Broad, enterprise-level integration	Deep, enterprise-wide integration
Role in Business	Peripheral, add-on tool	Important, but one part of the technological toolkit	Central to all business operations and strategies
Impact on Business Model	No fundamental change; digitization of processes	Enhances operations but structurally similar	Transforms business models and strategies
Feedback & Learning	Limited or no feedback loops	Some feedback for continuous improvement	Continuous, integrated feedback and adaptation
Data Integration	Isolated from other systems and data sources	Greater integration but not fully seamless	Seamless and enterprise-wide data integration
Culture and Mindset	Efficiency-focused	Efficiency and enhancement focused	Innovation and growth-focused
Technology Integration	Minimal, often in silos	More integrated, still some silos	Fully integrated across all systems
Innovation	Limited, mostly incremental	Incremental and some transformative	Highly transformative and generative
Operational Impact	Automates manual work for cost savings	Improves efficiency and customer experience	Drives new business models, market dominance

Table 2-1: The Comparison of Three AI Paradigms

This comparative analysis not only illustrates the evolving role of AI in business but also underscores the strategic considerations companies must navigate as they move along this spectrum. Understanding where a company currently stands in this continuum and where it aspires to be is crucial for effectively leveraging AI to achieve competitive advantage and drive innovation in today's rapidly changing business environment.

AI-Native: Redefining Business in the AI Era

The transition to AI-Native integration represents a watershed moment in the annals of business, comparable to the revolutionary impact of electricity and the advent of computing. This paradigm shift is not merely about adopting new technology; it's a comprehensive reimagining of business operations, innovation, and competitive strategies.

AI as the New Business DNA

In AI-Native organizations, AI's role is as intrinsic as DNA in living organisms, driving every facet of the business. This deep integration transforms the way companies formulate strategies, make decisions, and execute operations. AI-Native businesses are characterized by their extraordinary efficiency, adaptability, and capacity for innovation. They possess the foresight to anticipate market trends, the insight to make data-driven decisions, and the ability to offer exceptionally personalized customer experiences. In this context, AI becomes more than just a technological asset; it is the very essence of a company's identity, a fundamental characteristic that distinguishes it in a tech-centric marketplace.

The Evolutionary Leap with AI-Native

Adopting an AI-Native model marks a significant evolutionary step, transcending the realm of digital transformation to establish new benchmarks in business leadership and market redefinition. This shift is about proactively shaping the future of business, leveraging the full spectrum of AI's capabilities. As we move into an era dominated by Generative AI (GenAI), the competitive landscape will increasingly be defined by the ability of organizations to fully assimilate and harness AI's transformative potential.

A Clarion Call for Business Transformation

We stand at the threshold of an era where Generative AI is redefining industry norms, strategic paradigms, and competitive dynamics. The future will favor those who embrace this shift, embedding GenAI into the core of their business processes, thereby becoming more responsive, innovative, and customer-focused. This moment calls for decisive action - to embark on a transformative journey towards becoming an AI-Native enterprise. The imperative is clear: the future of business is inextricably linked to the mastery of Generative AI, making now the optimal time to embrace this groundbreaking shift. In the AI-Native future, the success of businesses will hinge on their ability to adapt, innovate, and redefine their operations in alignment with the advanced capabilities of AI.

Part Two: Creating a Compelling AI Vision and Winning Strategy

ArgoLong Publishing

Chapter 3

AI Strategy to Win

"To win by strategy is no less the role of a general than to win by arms." — Julius Caesar

T he alarm sounds jolting you awake. As your feet hit the floor, the smart room recognizes your wake-up routine and begins brewing your coffee while opening the blinds to gentle sunrise. Your digital assistant reads off your morning schedule, GPS-enabled smart shoes map the fastest route through your commute, and an autonomous car awaits to safely navigate rush hour as you catch up on email. AI sprinkles magic throughout your routine.

Yet this common vignette set in the near future starkly contrasts the state of most enterprises today, where AI largely remains siloed dust in the attic rather than golden innovation threaded through the fabric of operations and strategy. Static business models decay against unrelenting waves of disruption enabled by these exponential technologies. Survival hangs on building dynamic capabilities in AI to meet accelerating pace of change across technology, culture and globalization.

Mid-20th century business guru Peter Drucker famously declared that *"innovation is the specific tool of entrepreneurs, the means by which they exploit change as an opportunity"* and delight customers with things they never knew they needed. Today, AI empowers entrepreneurs large and small to reinvent entire ecosystems nearly overnight. Where once banking meant tellers and branches, now users transfer billions via mobile apps, VR enables attending far away events as if present, social networks rise and fall with public favor.

Yet with each leap into scarcely explored terrain, the margin for existential risk also compounds from unintended consequences of such powerful innovations. Like splitting the atom unleashing both terrible weapons and abundant electricity to serve mankind, responsible governance and ethics in applying AI introduces new realms for harm or help. The stakes have never been higher to deliberate carefully how we architect the operational logic of the modern world. Foundational business thinkers like Drucker could little conceive the tectonic implications of AI today. But the timeless wisdom to guide change rather than be changed holds firm.

AI Vision: The North Star Lighting An Uncharted Journey

Each monumental journey into unknown realms is anchored by a compelling vision, serving as a beacon that guides teams toward their lofty goals. Echoing the sentiments of the legendary explorer Ernest Shackleton during his early 20th-century Antarctic expedition, *"An enterprise is only as noble as its ambitions."* Our vision for AI over the next decade is to set high yet attainable goals, charting a course for both ambitious and grounded aspirations.

The most impactful technological transformations are those that focus not merely on the capabilities they provide but on the empowering possibilities they unlock for enhancing human life, freedom, and the pursuit of meaningful goals. Just as electricity led to inventions and infrastructure that significantly boosted productivity, connectivity, health, entertainment, and access to knowledge, the combustion engine revolutionized personal transportation and access. In a similar vein, AI and related technologies are poised to weave an intricate tapestry of new capabilities within our society over the next ten years. Our vision envisages AI subtly enhancing customer interactions through predictive personalization, while simultaneously breaking down barriers in products, services, education, and human dignity across various demographics. Imagine the smartphone revolution, which democratized information access over a decade, being compressed into three years through AI's accelerating power.

This AI-driven landscape demands not just incremental advancements but a complete overhaul of the enterprise. Transforming the organization to become **AI-native**, with intelligent interfaces, automation, and data-driven decision making at its core, necessitates a wholesale reimagining of every system, process, competency, and even our organizational culture. Anthropologist Margaret Mead astutely observed that a small group, courageous in its commitment to innovate and implement new paradigms, never fails to transform culture. Our vision calls for this kind of bold, transformative energy.

The initial step in any extensive journey may appear insignificant, but hindsight often reveals its crucial importance. Steady progress requires a vision that bolsters resolve through the inevitable false starts, delays, and fatigue, constantly reminding everyone involved why this particular venture is vital. Thus, our vision serves as both a guide and a reward, continually evolving and deepening with each lesson learned on this decades-long endeavor to institutionalize AI with responsible innovation at its heart. Every future is conceived in imagination; let our vision be the spark that lights the flame of ours, leading us toward a **Total Enterprise Reinvention** and the realization of an **AI-Native Enterprise**.

Total enterprise reinvention and the realization of an AI-Native Enterprise signify a profound transformation in how organizations operate in the era of artificial intelligence. This revolution transcends mere technological adoption, embedding AI into the core of business strategies, operations, and cultures. It involves reimagining traditional business processes and decision-making through the lens of data and AI capabilities, leading to more dynamic, adaptive, and efficient workflows. An AI-Native Enterprise not only automates routine tasks but also leverages AI for strategic insights, predicting market trends and customer behaviors. This transformation requires a cultural shift towards continuous learning and innovation, where employees are encouraged and equipped to work alongside AI technologies. It also necessitates a strong focus on customer-centric approaches, using AI to understand and anticipate customer needs better. The journey towards this new paradigm is marked by ethical considerations and sustainability in AI applications, ensuring that the integration of AI into businesses is responsible and beneficial for all stakeholders.

AI Predictions: 2024 and Beyond

As we look towards the future, the predictions from technology analysts at Gartner, Deloitte, McKinsey, and the World Economic Forum paint an evolving landscape shaped significantly by artificial intelligence.

Gartner's insights project a transformative impact on the workforce, with 30% of workers expected to use digital charisma filters and AI for enhanced communication by 2026. They also foresee AI's productivity value becoming a key economic indicator for countries by 2027, and predict a 26% increase in corporate investments to combat AI-enabled misinformation by 2028.

Deloitte's analysis highlights rapid advancements in machine learning, with a predicted 60% annual growth rate in AI patent applications globally. They emphasize AI's crucial role in driving sustainability, projecting a reduction in data center carbon emissions by over 35% in the coming years due to energy-efficient AI solutions.

McKinsey underscores AI's potential in addressing climate change and forecasts a staggering $13 trillion in new economic value from AI by 2030. They anticipate AI will significantly enhance functions like marketing, supply chain planning, and R&D within the next five years.

The World Economic Forum warns of the disruptive impact of AI on employment, predicting the elimination of 85 million jobs by 2025. This calls for effective workforce transitioning policies and a global reskilling effort for 1 billion people in response to AI's increasing role in business processes.

These predictions suggest that AI's influence will expand dramatically across economic, social, and environmental spheres by 2024 and beyond. The strategic integration of AI into operations, coupled with a focus on ethics and governance, is becoming increasingly important. As AI capabilities accelerate worldwide, the societal implications become more pronounced, necessitating timely actions from leaders to develop responsible AI strategies for their organizations.

AI Strategy Framework

n the rapidly evolving landscape of artificial intelligence, developing a cohesive and effective AI strategy is crucial for organizations seeking to leverage this transformative technology. We develop a comprehensive AI strategy framework designed to guide businesses through the complexities of AI integration and maximization. This framework is built around six strategic pillars: Vision, Value, Cost, Risk, Adoption, and Transformation. This strategic framework aims to equip leaders with the insights and tools needed to navigate the AI landscape successfully, ensuring that their AI investments yield tangible and sustainable results.

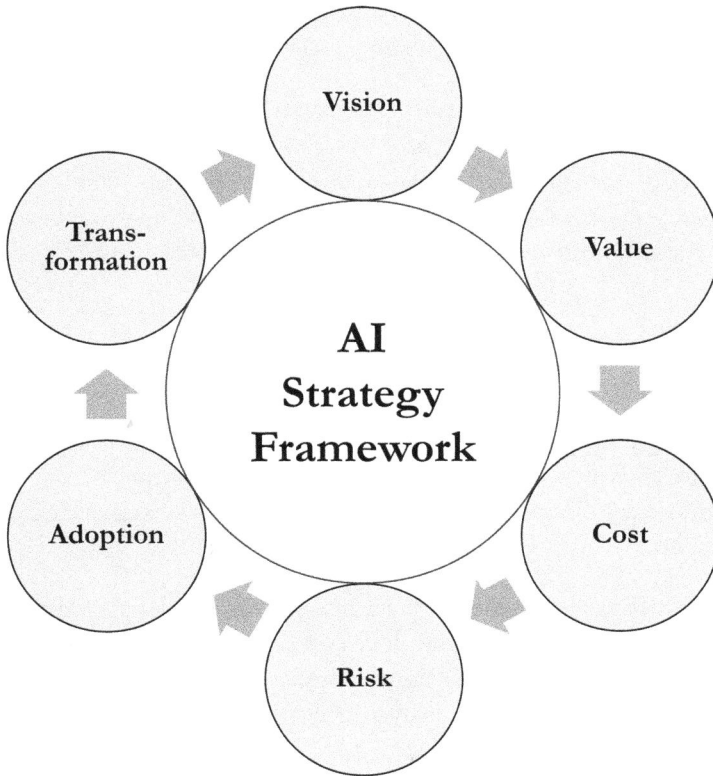

Figure 3-1: The AI Strategy Framework with Six Strategic Pillars

1. Vision: Total Enterprise Reinvention into an AI-Native Organization

Our 5-year AI vision acts as the north star - a fixed inspiration perpetually guiding teams towards a future in which AI dynamically advances organizational capabilities, processes and market leadership. This long-term conceptual focus promotes sustained investments immune to near-term financial pressures. We will craft a vivid, emotionally compelling vision statement bringing to life our AI-powered organization in 2030.

Enabling Total Enterprise Reinvention as an AI-Native Innovator

Specifically, our vision focuses on enabling the total enterprise reinvention and transformation required to become an AI-native organization. In an AI-native enterprise, AI is not bolted onto traditional business infrastructure and processes. Rather AI capabilities fundamentally enable key decisions, power product/service innovations, drive customer channels, personalize engagement and inform strategic market positioning through integrated intelligence.

Realizing this AI-native vision requires reimagining nearly all aspects of the company from the ground up, examining how generative AI can elevate efficiencies while also opening up new value propositions and revenue channels. This mandates reinventing roles, processes, systems and even organizational values from first principles optimized for symbiosis with AI.

Strategic Alignment with Corporate Vision

While a bold AI vision suggests an ambitious destination for organizational transformation, pragmatic progress requires tight alignment to the corporate vision and strategy. Our corporate vision outlines an overarching worldview on the desired future state and competitive positioning of the company as well as core values that guide decisions.

Effective AI vision complements this by painting an inspirational picture of how exponential technologies can enrich products, services, business models and customer experiences to accelerate achieving the corporate vision. Corporate strategy evolves the vision while AI strategy turbocharges execution of that evolving vision.

Centering Human Values

While technologies inevitably evolve, vision persists by tying AI outcomes back to enduring human values our customers share - health, connection, personal growth, education, creativity and living richly. Our vision therefore paints an emotionally compelling picture of an AI-powered future that uses exponential technologies not as an end, but as a means to expand prosperity, fairness and human dignity across the markets we serve.

With clarity on this north star guiding total enterprise reinvention as an AI-native digital innovator, leadership can assess which new AI capabilities and emerging technologies to prioritize investing in over milestone time horizons while benchmarking progress through transformation KPIs. Even amidst perpetual change, vision centered on improving lives endures as true north.

2. Value: Targeting High ROI and ROE

If vision lights the path, value creation fuels the journey at each step. Our strategy targets high return on investment (ROI) and return on experience (ROE) AI use cases that enhance productivity, decision making, customer intimacy and new offerings. The Value pillar involves a rigorous evaluation of potential generative AI use cases across three crucial dimensions to quantify and communicate the value generation for key stakeholders:

Productivity Gains

We assess opportunities for enhancing efficiency and automating tasks, such as using generative AI for social media posts, email campaigns, or document templates. Estimating cost savings from improved output quality, increased volume, reduced manual effort, and error reduction is conducted by comparing historical metrics before and after generative AI adoption. Processes consuming substantial employee time on low cognitive value tasks are highlighted showing automation potential to refocus efforts on creative and strategic activities.

Innovative Capabilities

Use cases offering unique capabilities that could disrupt market norms or enable new business models are investigated, showcasing how generative AI can drive innovation. Success stories from early adopters quantify the competitive advantage gained, drawing parallels with past technological waves like business intelligence analytics. Articulating long-term transformation possibilities in products, services, and customer experiences focuses on ambitious yet feasible scenarios.

Total Experience Enhancement

We delve into the concept of **Total Experience** (TX) by evaluating generative AI use cases that holistically enhance experiences for employees, customers, and stakeholders. Focus areas include generative AI-driven tools that empower employees, leading to increased productivity, satisfaction, and creativity, thereby positively impacting the overall workplace environment. Examining customer-facing applications like personalized interaction systems and predictive services highlights more intuitive and tailored user experiences. Assessing the interconnected impact of improved employee and customer experiences on overall business health translates experience metrics into tangible business outcomes and strategic advantages.

Evaluating this trio of value dimensions — productivity, innovation, experience — facilitates an objective view of generative AI's total value potential, allowing calibration of investments to priority areas while building an enterprise-wide perspective. This assessment methodology also structures compelling business cases using relevant metrics and market benchmarks tailored to diverse stakeholders.

3. Cost: Navigating Financial Implications

The Costs pillar involves comprehensive modeling and projection of the total cost of ownership (TCO) for developing, deploying and operating generative AI capabilities. Evaluating nine key expense categories across the solution life cycle prevents unanticipated overruns while empowering decision-making through realistic budget planning:

1. **GenAI Tools & Platform Access Costs:** Includes licensing fees for tools like ChatGPT Plus, subscription costs for commercial AI platforms (OpenAI, Anthropic, Cohere, AI21 Labs), and variable API usage fees for accessing external datasets and models.

2. **Prompt Engineering Costs:** Encompasses investments in prompt creation tools and template libraries, costs for hiring specialized engineers, and training expenses for employees to develop proficiency in prompt engineering.

3. **Inference Costs:** Covers costs related to the usage of Large Language Models (LLMs) such as GPT-4, including token-based pricing for model input and output, and the infrastructure and energy costs for running high-performance servers.

4. **Fine-Tuning Costs:** Involves expenses based on the size and complexity of the model, the volume of data used for fine-tuning, and the number of training epochs required, with potential investments in cost-effective fine-tuning platforms.

5. **Infrastructure Costs:** Comprises Cloud hosting expenses, costs for integrating GenAI with legacy systems, and investment in computational resources necessary for running AI models.

6. **Data Management Costs:** Includes expenses for data storage upgrades, costs associated with data engineering tools and processes, licensing fees for external data, and human review expenses for data annotation.

7. **Operations Costs:** Consists of continuous learning and retraining expenses, monitoring and maintenance costs, and investments in MLOps and AIOps for optimizing AI operations.

8. **AI Regulations Compliance Costs:** Covers expenses for ensuring AI systems' transparency, fairness, and security, costs for legal and ethical

compliance measures, and investments in sustainable and human-centric AI development.

9. **Talent Costs:** Encompasses hiring AI leaders, balancing immediate and long-term talent needs, costs for leadership and cultural development, expenses for upskilling and reskilling programs, and adapting to remote work considerations for AI professionals.

Modeling TCO holistically rather than just initial price tag allows prudent budgeting and scaling. Distilling complex projections through simplified cost groupings makes communications with stakeholders more accessible while highlighting priority focus areas to drive efficiency. Keeping tabs on all contributors to generative AI's costs is key to summiting value heights responsibly.

4. Risk: Addressing New Challenges

The Risks pillar is critical in the responsible adoption of generative AI, focusing on a thorough evaluation and mitigation of potential challenges in quality, security, privacy, ethics, and compliance.

- **Output Quality Risks:** Addressing issues like factual inaccuracies, biases, or hallucinated content through continuous monitoring, input validation, and human-in-the-loop reviews is crucial to ensure the reliability of generative AI outputs.

- **Data Security Risks:** Protecting sensitive datasets used in model training is essential to prevent data breaches or intellectual property theft. Employing encryption, stringent access controls, and data masking techniques are key safeguards.

- **Privacy Risks:** Despite anonymization, the risk of attribute leakage can lead to privacy rights violations. Implementing differential privacy methods, anonymization checks, and consent procedures are necessary for privacy compliance.

- **Bias and Fairness Risks:** Mitigating historical biases in training data is vital to avoid discriminatory AI outputs, requiring proactive bias testing and mitigation processes.

- **Transparency and Explainability Risks:** Ensuring model transparency is essential for identifying failure points and building trust. Integrating explainability methods and maintaining documentation standards help in mitigating these risks.

- **Misuse and Harms Risks:** Preventing the misuse of generative models to spread misinformation or facilitate adversarial attacks involves establishing oversight procedures and operational guardrails.

- **Compliance Risks:** Navigating regulatory policies like GDPR or emerging AI laws requires continuous auditing and adaptation to avoid financial penalties or licensing issues.

By proactively identifying and mitigating these risks, organizations can embed necessary safeguards into their AI strategies, ensuring data integrity, model governance, ethical AI practices, and compliance. This approach also helps in assessing insurance needs and preparing for worst-case scenarios, thus fostering resilience and ethical progress in AI applications.

5. Adoption: Evaluating Organizational Readiness

The adoption pillar is centered on a comprehensive audit of organizational readiness, identifying potential bottlenecks in talent, infrastructure, data, and governance that could impact successful AI adoption and scaling. A thorough assessment of the organization's maturity level is crucial to bridge existing gaps.

- **Talent Readiness:** Conduct surveys to assess current skills in machine learning engineering, data science, and prompt engineering, identifying talent deficiencies that might hinder AI adoption.

- **Leadership Alignment:** Engage with executives and managers to understand their perceptions, concerns, and readiness for AI-driven change, ensuring alignment at the leadership level for cultural transformation.

- **Data Readiness:** Examine the data inventory in relation to specific use case requirements, focusing on aspects like volume, labeling, privacy, lineage, and lifecycle management.

- **Infrastructure Readiness:** Assess the existing technology infrastructure's capability to support the accelerated experimentation and deployment demands of AI systems.

- **Governance Readiness:** Review current practices around quality, trust, and compliance to refine policies in critical AI areas such as accountability, transparency, and fairness.

- **Evolving Maturity:** In the face of rapid AI advancements, continuously build and reassess capabilities to keep pace with the exponential shifts in technology.

Conducting honest evaluations of readiness, avoiding overconfidence bias, is key to pragmatic planning, effective talent development, and fostering a culture conducive to transformative change. Realistically appraising an organization's preparedness for generative AI is as crucial as a climber's realistic assessment of their abilities before attempting a challenging summit, ensuring successful and scalable AI adoption.

6. Transformation: Steering A New Journey of AI-Native Enterprise

Lastly, the transformation pillar emphasizes the need for cultural and operational shifts to seamlessly integrate generative AI into business practices. Effectively managing this change is crucial for the successful realization of adoption outcomes.

- **Cultural Alignment**: Cultivate a culture receptive to AI-driven transformation by fostering understanding and enthusiasm. Use transparent communication and showcase quick wins to build organizational confidence in AI capabilities.

- **Change Management:** Establish clear transition milestones in line with the AI capability roadmap. Support teams in navigating interim workflow adaptations and new tool integrations, leveraging best practices from digital transformation experiences.

- **Skills Development:** Conduct a comprehensive skills gap analysis and roll out tailored training programs, both self-paced and cohort-based, to enhance knowledge in areas like data science, AI, and ethics.

- **New Role Creation:** Innovate in workforce structuring by designing new roles such as prompt engineers, AI trainers, and trust assessors, which blend existing skills in novel ways to facilitate AI advancement.

- **Process Re-engineering:** Reassess and redesign processes with an eye toward optimizing human-AI collaboration, moving beyond the mere automation of existing methods.

- **Governance Realignment:** Adapt governance frameworks to ensure accountability and transparency in AI systems. Implement multi-disciplinary councils, involving ethicists, to uphold AI quality and ethical standards.

By placing equal emphasis on people and technology transformation, organizations can effectively navigate from conceptualization to impactful implementation of GenAI. This journey, led with empathy, clarity, and transformational leadership, is essential to foster employee engagement and active participation in shaping an AI-driven future. As generative AI redefines the workplace, leaders must leverage digital transformation insights, ensuring a seamless transition into an AI-enhanced future. Equipped with a a digital mindset, leaders are instrumental in creating an environment where employees are not just involved but also positively impacted by generative AI, making them active beneficiaries of this technological evolution.

In conclusion, the journey towards successfully integrating and leveraging AI in your organization is a multifaceted endeavor that requires a holistic approach. The six-pillar framework of Vision, Value, Cost, Risk, Adoption, and Transformation offers a structured and strategic pathway to navigate this complex landscape. By carefully considering each of these aspects, organizations can not only harness the power of AI to achieve immediate goals but also lay the groundwork for sustained innovation and growth. With a well-thought-out AI strategy, businesses can unlock the full potential of artificial intelligence, turning it into a key driver of success and competitive advantage in the digital era.

AI-Native Thinking: Reshaping Strategic Business Choices

In the realm of business strategy, the "**Playing to Win**" framework, conceptualized by A.G. Lafley and Roger Martin, has been a cornerstone for organizations seeking a competitive edge. This approach emphasizes making deliberate choices in five key areas: Winning Aspiration, Where to Play, How to Win, Critical Capabilities, and Management Systems. Each choice is interdependent, guiding companies in defining their purpose, selecting their markets, determining their approach to success, building necessary capabilities, and managing operations effectively.

However, in the wake of the AI revolution, a new dimension has emerged - AI-native thinking. This paradigm shift reinterprets the "Playing to Win" framework through the lens of AI, transforming traditional strategic models into dynamic, AI-centric strategies. As businesses transition into AI-native enterprises, they must reassess and redefine their approach to these five strategic choices to capitalize on the potential of AI technologies.

AI-native thinking is not just about incorporating AI tools into existing strategies; it's about reimagining every aspect of business strategy with AI at the core. It demands a radical rethink of how a business defines its winning aspiration, chooses its markets, establishes its competitive advantage, develops its capabilities, and manages its systems. This approach goes beyond mere adaptation to AI; it's about leveraging AI as the driving force behind innovation, operational efficiency, and market leadership.

In this context, the traditional "Playing to Win" framework evolves. The winning aspiration becomes more dynamic, focusing on AI-driven growth and innovation. 'Where to Play' and 'How to Win' choices expand to include AI-powered markets and strategies. Critical capabilities must now encompass AI-centric skills and infrastructure, and management systems need to be agile and adaptable to keep pace with AI advancements.

This chapter aims to bridge the gap between traditional strategic planning and the emerging AI-native approach. As we explore each strategic choice in the following sections, we'll see how AI-native thinking not only aligns with but also enhances the principles of the "Playing to Win" framework, offering businesses a roadmap to thrive in the digital age.

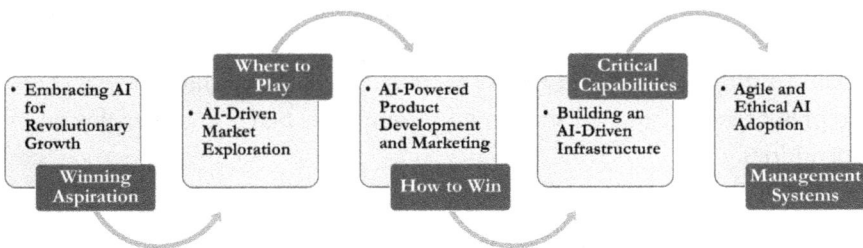

Figure 3-2: The AI-Native Thinking Reshapes the Five "Playing to Win" Strategic Business Choices

1. Winning Aspiration: Embracing AI for Revolutionary Growth

In an AI-native enterprise, the 'Winning Aspiration' transcends traditional goals of market leadership or financial metrics. This strategic choice now pivots around harnessing AI for groundbreaking innovation and growth. Success in this realm is redefined: it's not merely about dominating existing markets but about creating new ones through AI-driven transformation. The aspiration becomes a vision where AI is the catalyst for uncovering novel business opportunities and redefining industry norms.

AI-Driven Business Model Innovation

Consider how AI can revolutionize business models. For instance, a manufacturing firm traditionally focused on production efficiency might evolve into a leader in smart manufacturing. By integrating AI for predictive maintenance and supply chain optimization, it not only enhances its operational efficiency but also pioneers new standards in the industry. Similarly, in sectors like retail, AI can transform customer experiences. Retailers leveraging AI for personalized shopping experiences or virtual try-on solutions are not just improving service but are also reimagining the retail landscape.

AI as the Gateway to New Markets

The AI-native winning aspiration also involves identifying and seizing new market opportunities through AI. Financial services, for example, can be transformed with AI applications in real-time risk assessment, leading to innovative products in micro-lending or tailored insurance services. These are not incremental changes but represent a fundamental shift in how companies perceive and exploit market opportunities.

Challenges in Operationalizing an AI-Native Vision

Operationalizing this AI-centric winning aspiration involves more than technological adoption. It requires cultivating a culture that embraces innovation and views AI as a core component of business strategy. Investing in AI talent and infrastructure is crucial, as is the commitment to continuous learning and adaptation in the face of evolving AI technologies.

However, this journey is not without its challenges. Ethical considerations, data privacy, and responsible AI use are critical issues that companies must address.

Moreover, integrating AI into the core strategy entails significant investments in not just technology but also in transforming organizational culture and mindset.

In conclusion, the AI-native winning aspiration is about envisioning a future where AI is not just an enabler but the driving force of business strategy. It's about leveraging AI to unlock unprecedented growth, innovate business models, and redefine customer engagement. This strategic shift is pivotal for companies aiming to lead in the rapidly evolving digital landscape.

2. Where to Play: AI-Driven Market Exploration

The strategic choice of 'Where to Play' fundamentally shifts in an AI-native enterprise. Traditionally, this choice involves selecting market segments, geographies, and customer bases where a company competes. However, in the context of AI-driven strategy, this choice expands dramatically. AI enables companies to explore and enter markets that were previously inaccessible or nonexistent, effectively redrawing the boundaries of their business landscape.

Exploring New Market Territories with AI

AI's ability to analyze vast datasets and generate insights allows companies to identify emerging trends and underserved market niches. For instance, an AI-native company might discover a previously unnoticed demographic with specific needs or a geographic market ripe for digital services. AI-driven market exploration is about leveraging these insights to venture into new territories, whether they are adjacent markets or entirely new sectors.

AI in Creating New Customer Value Propositions

In addition to identifying new markets, AI enables the creation of innovative value propositions. Companies can use AI to tailor products and services to specific customer needs, offering a level of personalization that sets them apart in the marketplace. This approach not only enhances customer engagement but also opens up new avenues for revenue generation.

The Dynamic Nature of AI in Market Strategy

The dynamism of AI extends to how companies view their competitive landscape. AI-native thinking encourages continuous exploration and adaptability. As AI technologies evolve, so do the opportunities for market expansion and diversification. Companies can quickly prototype and test new products or services using AI, reducing the time and cost traditionally associated with market exploration.

Challenges in AI-Driven Market Exploration

While AI offers remarkable opportunities for market exploration, it also presents challenges. Companies must navigate the complexities of data-driven decision-making, ensuring they have robust data governance and analytics capabilities. Additionally, there's a need to align AI-driven market choices with the overall strategic vision of the company, ensuring that these ventures are sustainable and in line with long-term goals.

'Where to Play' in an AI-native enterprise is not just a choice but a continuous exploration. AI not only allows companies to identify and enter new markets but also to create unique value propositions and stay agile in a rapidly changing business environment. This strategic choice becomes an ongoing process of discovery, driven by the transformative power of AI.

3. How to Win: AI-Powered Product Development and Marketing

In the AI-native enterprise, 'How to Win' is fundamentally transformed. This strategic choice, traditionally about defining the unique selling propositions and key differentiators, is reimagined with AI at its core. AI is not just a tool in the arsenal; it becomes the central element in developing competitive strategies, influencing both product development and marketing.

AI-Driven Product Development

AI revolutionizes the product development process. It enables companies to leverage data-driven insights for innovation, creating products that are not only responsive to current market needs but also predictive of future trends. AI algorithms can analyze customer feedback, market trends, and operational data to suggest improvements, identify new product opportunities, and even predict potential market successes.

For example, in the automotive industry, AI can be used to design cars with advanced safety features or eco-friendly innovations, addressing both current consumer demands and regulatory trends. In consumer electronics, AI can guide the development of smart home devices that cater to the increasingly connected and automated lifestyles of users.

Personalization and Precision in Marketing

AI's impact on marketing is equally transformative. Personalization becomes the watchword, with AI analyzing consumer behavior to tailor marketing messages to individual preferences and needs. This level of personalization enhances customer engagement and loyalty, as consumers feel more understood and valued.

Moreover, AI enables precision marketing. Companies can use AI to optimize their marketing spend, targeting the right consumers at the right time through the right channels. AI-driven analytics can predict which marketing strategies are likely to be most effective, reducing the trial-and-error aspect of marketing campaigns.

Continuous Innovation and Responsiveness

AI fosters a culture of continuous innovation and responsiveness. In a fast-paced market, companies can quickly adapt their products and marketing strategies in response to real-time data and insights provided by AI. This agility is crucial in maintaining a competitive edge.

Challenges of AI in Product Development and Marketing

Despite these advantages, integrating AI into product development and marketing is not without challenges. Companies must ensure they have the necessary infrastructure and talent to leverage AI effectively. There's also the need to maintain a balance between AI-driven decisions and human judgment, especially in understanding the nuances of customer behavior and market dynamics.

'How to Win' in an AI-native landscape is about harnessing the power of AI to drive innovative product development and highly personalized, precision marketing. It's a shift from traditional competitive strategies to dynamic, data-driven approaches that are continuously evolving in line with technological advancements and market trends.

4. Critical Capabilities: Building an AI-Driven Infrastructure

In an AI-native enterprise, the 'Critical Capabilities' element of strategy shifts focus towards building a robust AI-driven infrastructure. This infrastructure is not just about the technology itself but encompasses the systems, processes, and talent that enable an organization to fully leverage AI for strategic advantage.

Creating a Comprehensive AI Ecosystem

The heart of this infrastructure is a comprehensive AI ecosystem. This includes advanced data analytics platforms, machine learning algorithms, cloud computing resources, and edge computing capabilities. Such an ecosystem allows for the efficient processing and analysis of large volumes of data, essential for AI-driven decision-making and innovation.

Developing Proprietary AI Systems and Tools

Developing proprietary AI systems and tools is another critical aspect. These tailored solutions can provide a competitive edge by addressing specific organizational needs and challenges. For instance, a retail company might develop an AI system for optimizing inventory management, reducing waste and increasing efficiency.

Fostering AI Talent and Skills

Equally important is the cultivation of AI talent. This involves not only hiring experts in AI and data science but also upskilling existing staff. Providing training and resources to understand and work alongside AI technologies ensures that the entire organization can contribute to and benefit from AI initiatives.

Integrating AI into Operational Processes

Integrating AI into operational processes is crucial for realizing its full potential. This means embedding AI into various facets of the business, from customer service to supply chain management. For example, AI can be used to predict supply chain disruptions and automatically adjust orders and inventory levels, ensuring business continuity.

Challenges of Building an AI-Driven Infrastructure

Building an AI-driven infrastructure is a significant undertaking. It requires substantial investment in technology and human capital. There are also challenges related to data privacy and security, ethical considerations in AI deployment, and ensuring the interoperability of AI systems with existing IT infrastructure.

The 'Critical Capabilities' aspect of AI-native strategy revolves around establishing a solid foundation of AI infrastructure, systems, and talent. This infrastructure is the bedrock upon which AI-driven innovation and strategy are built, enabling organizations to harness the full power of AI for competitive advantage. Overcoming the challenges associated with this endeavor is key to realizing the transformative potential of AI in business.

5. Management Systems: Agile and Ethical AI Adoption

In an AI-native enterprise, the strategic element of 'Management Systems' undergoes a significant transformation. It's no longer just about traditional management practices but about creating systems that support agile and ethical adoption of AI technologies. This involves rethinking organizational structures, decision-making processes, and cultural norms to align with the dynamic nature of AI.

Agile Frameworks for AI Integration

Agility becomes a key component of management systems in the AI era. Organizations must adopt flexible and responsive management frameworks that can quickly adapt to the rapid developments in AI technology. This means moving away from rigid hierarchical structures to more fluid and cross-functional teams. It also involves adopting project management methodologies like Scrum or Kanban, which are well-suited to the iterative and experimental nature of AI projects.

Ethical Considerations in AI Deployment

Ethical deployment of AI is a critical aspect of management systems. This involves establishing guidelines and practices to ensure that AI technologies are used responsibly. Companies must address issues such as data privacy, bias in AI algorithms, and the ethical implications of AI decisions. Creating a governance framework that oversees the ethical use of AI ensures that the company's AI initiatives align with societal values and regulations.

Cultivating a Culture of Continuous Learning

An AI-native enterprise needs a culture that values continuous learning and adaptability. AI technologies are continually evolving, and organizations need to ensure their employees are equipped to keep pace. This involves not only technical training but also fostering an environment that encourages experimentation and learning from failures. A culture that embraces change and innovation is essential for the successful adoption of AI.

Challenges in Implementing Agile and Ethical AI Systems

Implementing agile and ethical management systems for AI is not without its challenges. It requires a fundamental shift in how organizations operate and make decisions. There are also challenges in ensuring that all employees, from leadership to frontline staff, understand and are aligned with these new approaches. Additionally, balancing the need for rapid AI development with ethical considerations and regulatory compliance can be complex.

The 'Management Systems' component of an AI-native strategy is about creating an organizational environment that supports the agile and ethical deployment of AI. It requires flexible management structures, a strong ethical framework, and a culture of continuous learning and adaptation. Overcoming the challenges in this area is crucial for organizations to fully leverage the benefits of AI and ensure their initiatives are sustainable and responsible.

In conclusion, the journey towards becoming an AI-native enterprise, as explored through the transformation of the five strategic choices – Winning Aspiration, Where to Play, How to Win, Critical Capabilities, and Management Systems – represents a pivotal shift in business strategy in the AI era. This evolution is not just about technology adoption, but a profound rethinking of business models and practices. While the path is fraught with challenges, including significant investments, ethical considerations, and cultural transformations, the rewards promise immense competitive advantages. Companies that successfully integrate AI into their core strategy will unlock new growth opportunities, innovate at unprecedented speeds, and lead in an increasingly digital world. As we embrace this future, AI-native thinking stands not just as a technological imperative, but as a strategic cornerstone for businesses aiming to thrive in the evolving landscape of the digital age.

A Lean Approach to AI Strategic Planning

The strategic planning process for AI initiatives demands a nimble, focused, and responsive framework. By adopting a lean approach, organizations can craft a strategy that is both agile and robust, enabling quick adaptation to the rapidly evolving AI landscape. Here is an adapted 12-step process that embodies a lean philosophy in AI strategic planning:

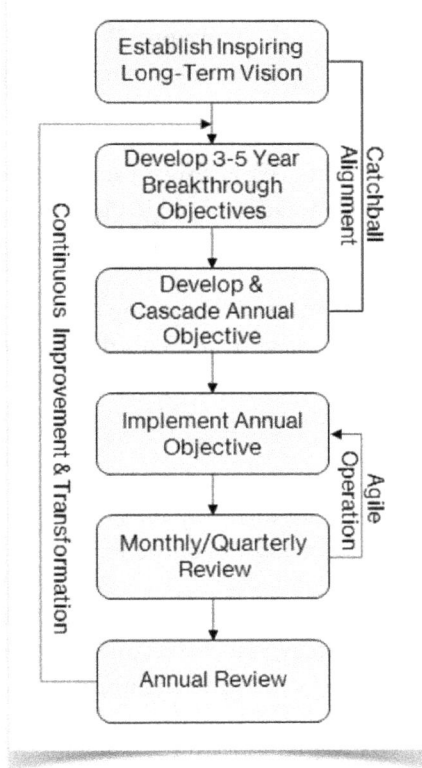

Figure 3-3: AI Strategic Planning Process (Adapted from Hoshin Kanri Planning)

1. **Streamlining the Process**: Refine the traditional Hoshin Kanri Planning Process to suit the fast-paced nature of AI development. This lean version eliminates waste, focuses on value, and accelerates decision-making.

2. **Visionary Alignment**: Articulate a compelling long-term vision for the AI Tech organization that is in harmony with the corporate strategy. This vision should inspire innovation and guide strategic decisions.

3. **Bold Goals with Precision**: Set forth 3-Year Big Hairy Audacious Goals (BHAGs) for the AI Tech organization utilizing the Objectives and Key Results (OKRs) style. This combination ensures that goals are not only ambitious but also measurable and tightly aligned with strategic priorities.

4. **Foundation of Strategy Pillars**: Identify and establish the main pillars of the AI strategy, along with the business value drivers. These pillars will support the overarching strategy and ensure that every tactical move drives towards the end goals.

5. **Empowering Teams with Tools**: Equip teams with strategic planning guidance and templates well in advance. Providing the right tools fosters a sense of ownership and preparedness as teams contribute to the strategy sessions.

6. **Collaborative Planning Workshop**: Facilitate a one-day onsite workshop to engage in collaborative strategic planning. This session is pivotal in integrating diverse insights and forging a shared strategic path.

7. **Baseline Strategy Formation**: Following the workshop, draft a baseline version of the 3-Year Strategy Plan. This document will encapsulate the collective intelligence and learning from the workshop, serving as a foundation for further refinement.

8. **Securing Team Commitment**: Conduct in-depth discussions with functional teams to secure their buy-in. Winning team support is critical for the successful implementation of the strategy.

9. **Iterative Strategy Enhancement**: Implement two half-day virtual sessions to refine the strategy plan based on team feedback. This iterative process allows the strategy to evolve and improve continuously.

10. **Stabilizing the Strategic Vision**: Consolidate the feedback to create a stable version of the 3-Year Strategy Plan that includes a Strategic Portfolio, Risk Dashboard, and OKRs Dashboard. These tools provide clarity and facilitate ongoing monitoring and adjustment.

11. **Broad Communication**: Publish and broadly communicate the Strategy Plan across the organization. Transparent communication ensures that all stakeholders are aligned and engaged with the strategic direction.

12. **Dynamic Strategy Updates**: Maintain the flexibility to update the Strategy Plan quarterly or as required. A lean approach to strategic planning is inherently dynamic, accommodating new insights and adapting to changes with agility.

In embracing "A Lean Approach to AI Strategic Planning", organizations can cultivate a strategic mindset that is as dynamic and forward-thinking as the AI field itself. This 12-step process is but an example, a template from which businesses can begin to craft strategies that are both resilient and adaptable to the unpredictable tides of technological advancement. While this lean methodology offers a pathway to agile innovation, it is by no means the only route to strategic success in AI. Organizations may find value in exploring a variety of strategic planning methods, each offering distinct perspectives and tools. Ultimately, the best approach is one that aligns with the unique culture, goals, and challenges of the organization, fostering a strategic plan that not only guides but also grows with the company in the ever-evolving landscape of artificial intelligence.

Conclusion

As we close the chapter on "AI Strategy to Win", we stand at the crossroads of anticipation and action. We have navigated through the predictions for AI in 2024 and beyond, setting the stage for a future where artificial intelligence is not just an enabler but the cornerstone of strategic innovation. We've introduced an AI Strategy Framework underpinned by six strategic pillars and embraced AI-native thinking to reshape the classical "Playing to Win" business choices. We also charted a course for a lean approach to strategic planning, tailored for the swift currents of the AI revolution.

Looking ahead, our journey into the vast and intricate realm of AI strategy continues. The next six chapters will delve into each strategic pillar in meticulous detail, unpacking their significance, exploring their interconnections, and presenting a blueprint for their operationalization within the AI domain.

Chapter 4

AI Business Value Octagon

"When your values are clear to you, making decisions becomes easier." — Roy E. Disney

A s businesses navigate the evolving landscape of technology, Artificial Intelligence (AI) stands as a beacon of transformation and innovation. However, a significant challenge persists - the Value Gap in AI adoption. This gap represents the disconnect between the potential of AI and its actual realization in business strategies. Many leaders, while recognizing AI's transformative power, struggle to integrate it effectively into their business models, often due to a lack of understanding of its practical applications and strategic value. To bridge this gap, a deep understanding of AI's impact across industries and organizational functions is crucial.

Our comprehensive approach involves analyzing an array of business cases across various sectors and organizational functions, providing a clear picture of AI's diverse applications and their outcomes. This analysis forms the backbone of the 'Value Pillar' of AI strategy, a robust tool designed to guide leaders in aligning AI initiatives with their strategic objectives. By focusing on relevant and impactful AI applications, this framework steers organizations away from superficial enhancements towards significant, value-driven AI integration.

Incorporated into this approach is the 'AI Business Value Octagon' framework. This innovative model breaks down the multifaceted impact of AI into eight key dimensions, offering a holistic view of how AI can drive business value. It serves as an essential guide for prioritizing AI investments in areas that promise the highest returns, aligning with the organization's strategic goals.

AI Business Cases Across Industries

The advent of AI in the business world has ushered in a new era of operational efficiency and innovation, transcending traditional industry boundaries. This section delves into the multifaceted impact of AI across various sectors, drawing insights from specific industry examples to illustrate AI's transformative potential.

Industry	Value Highlighted	3 High-Value Applications
Life Sciences & Healthcare	▪ Accelerate innovation ▪ Strengthen oversight ▪ Boost efficiency	▪ Molecular modeling for drug discovery ▪ Ambient digital scribes ▪ Adverse event detection
Retail	▪ Boost revenue and loyalty ▪ Improve efficiency	▪ Enhanced search and recommendations ▪ Supply chain and merchandising optimization ▪ Conversational interfaces
Finance	▪ Boost revenue ▪ Mitigate risk ▪ Increase efficiency	▪ Synthetic data for credit risk models ▪ Transaction monitoring ▪ Portfolio rebalancing
Manufacturing	▪ Boost efficiency, quality and sustainability	▪ Equipment effectiveness optimization ▪ Rapid prototyping ▪ Intelligent robotic systems
Education	▪ Boost revenue ▪ Accelerate innovation ▪ Improve efficiency	▪ Personalized tutors, ▪ Learner matching and nudging ▪ Decision simulations

Table 4-1: The High-Value Business Cases Across Industries

Realizing the Business Value of Generative AI in Life Sciences and Healthcare

Across critical functions like drug discovery, patient care, and medical research, generative AI shows immense promise to accelerate innovation, strengthen oversight, and boost efficiency. As biopharma executives and healthcare administrators explore investments in this emerging capability, evaluating the potential business value across key use cases is imperative. Based on an assessment of numerous applications, three tiers of impact have emerged:

High-Value Applications: Several generative AI solutions stand out for their potential to expand treatment pipelines, improve clinician productivity, and increase safety. Molecular modeling and scientific data analysis promise to radically accelerate novel compound development. Meanwhile, ambient digital scribes and virtual care assistants have tremendous value for unlocking clinician capacity and healthcare access. Adverse event detection, clinical trial matching, and personalized care coordination additionally help strengthen quality, equity and outcomes.

Moderate-Value Applications: More incremental yet still important potential exists in areas like automated regulatory filings, patient education, and population health analysis. Generative AI's rapid content creation could also simplify policy documentation and patient-facing materials. Supply chain coordination, drug safety signaling, and diagnostics additionally warrant consideration.

Lower-Value Applications: While intriguing, emerging use cases around fully autonomous diagnostics, unsupervised lab procedure execution, and broad operational compliance may underwhelm in their near-term impact. Proving standalone value for these innovative but highly complex functions could be challenging, although they may complement broader digital transformation.

Realizing the Business Value of Generative AI in Retail

The advent of generative AI heralds a transformative era in retail, offering innovative solutions to enhance key functions such as search, personalization, planning, and customer service. As retailers consider investing in this burgeoning technology, a thorough evaluation of its potential business value across various applications is crucial. An in-depth analysis of numerous applications reveals a tiered impact structure:

High-Value Retail Applications: Several generative AI solutions stand out for their ability to boost revenue and loyalty or radically improve efficiency. Enhanced search and recommendations promise to increase conversion rates and average order values by helping customers find the right products. Meanwhile, optimizing supply chain and merchandising decisions through data synthesis unlocks major productivity gains. Conversational interfaces also build stronger customer relationships while slashing service costs.

Moderate-Value Use Cases: Social analytics, personalized marketing, and better demand forecasting offer less dramatic yet still important benefits. Generative AI's content creation talents additionally simplify promotional content production. Applications like customer order substitutions and workforce scheduling optimization also fall into this middle-tier based on their more incremental efficiency impacts.

Lower-Value Applications: Emerging use cases like immersive retail experiences, dynamic pricing models, and product co-creation with customers may underwhelm in their near-term value. While intriguing, most retailers will struggle to implement these capabilities at scale or realize major financial returns initially. Exceptions exist for categories like home goods where virtual design showcases provide differentiation.

Realizing the Business Value of Generative AI in Finance

Across banking, investments, and insurance, generative AI promises to reshape critical functions from customer service to compliance. As financial services firms explore investments in this emerging capability, evaluating the potential business value across key use cases is imperative. Based on an assessment of numerous applications, three tiers of impact have emerged:

High-Value Financial Services Applications: Several generative AI solutions stand out for their ability to boost revenue, mitigate risk, and radicalize efficiency. In banking, synthetic data promises to expand financial access and power more accurate credit risk models. Transaction monitoring and fraud prevention also see upside from process automation and pattern detection. In investments and insurance, advisor productivity surges using next-best-action recommendations while portfolio rebalancing and claims processing are streamlined.

Moderate-Value Use Cases: Personalized marketing, virtual assistants, customer churn predictions, and automated reporting offer less dramatic yet still important benefits. Generative AI's rapid content creation additionally simplifies regulatory reporting and client-facing documents. Workflow enhancements around account opening, servicing, and exception processing also warrant consideration.

Lower-Value Applications: While intriguing, emerging use cases around branch space design, customer prospecting, and compliance support may underwhelm in their near-term financial impact. Proving standalone value for these narrow functions could be challenging, although they may complement broader digital transformation.

Realizing the Business Value of Generative AI in Manufacturing

Across smart factories and digital supply chains, generative AI promises to reshape critical functions from production to maintenance. As manufacturers explore investments in this emerging capability, evaluating the potential business value across key use cases is imperative. Based on an assessment of numerous applications, three tiers of impact have emerged:

High-Value Manufacturing Applications: Several generative AI solutions stand out for their ability to boost efficiency, quality, and sustainability. By synthesizing insights across disparate datasets, factories can continuously optimize equipment effectiveness and identify process improvements. Meanwhile, rapid 3D visualization and design prototyping accelerate new product development and customization. Avatar brand representatives and intelligent robotic systems additionally open new market opportunities while enhancing safety.

Moderate-Value Use Cases: More incremental yet still important benefits come from AI-guided machine maintenance, real-time supply chain coordination, and personalized customer experiences. Generative AI also shows promise for automating software testing and documenting best practices. Workflow and data analysis enhancements similarly warrant consideration.

Lower-Value Applications: While intriguing, emerging use cases around operations dashboard creation, employee self-service tools, and branch space design may underwhelm in their near-term financial impact. Proving standalone value for these narrow functions could be challenging, although they may complement broader digital transformation.

Realizing the Business Value of Generative AI in Education

Across teaching, research, and administration, generative AI shows immense promise to enhance productivity and transform learning. As education institutions explore investments in this emerging capability, evaluating the potential business value across key use cases is critical. Based on an assessment of numerous applications, three tiers of impact have emerged:

High-Potential Education Applications: Several generative AI solutions stand out for their potential to boost revenue, accelerate innovation, and radicalize efficiency. Personalized tutors and teaching assistants could scale student support and engagement at lower cost. Meanwhile, optimized learner matching and nudging promises to increase enrollment yields, persistence, and alumni lifetime value. Accelerated research, content creation, and decision simulations additionally enable institutions to amplify their societal impact.

Moderate-Potential Use Cases: More incremental yet still important potential exists in areas like multilingual instruction, student services, and personalized career guidance. Generative AI's rapid content creation could also simplify regulatory reporting and student prospecting. Plagiarism detection and grading process enhancements similarly warrant consideration.

Lower-Potential Applications: While intriguing, emerging use cases around space utilization planning, fundraising targeting, and mental health interventions may underwhelm in their near-term potential. Proving value for these innovative but narrow functions could be challenging, although they may complement broader digital transformation.

AI Business Cases Across Organizational Functions

While generative AI holds promise across the enterprise, its applications and expected business value differ significantly between functions. IT plans to leverage the technology for improved software development, data analysis, and research. Marketing and sales see major upside in personalized content creation, customer service enhancements, and optimized prospecting.

Meanwhile, legal aims to transform document discovery and contract reviews. HR envisions more customized talent attraction and internal advising capabilities. The list continues with use cases in corporate strategy, finance, and beyond. Each function brings its own priorities centered around how AI can elevate processes, insights, and engagement.

Function	Value Highlighted	3 High-Value Applications
Executives & Boards	▪ Accelerate planning ▪ Strengthen controls ▪ Uncover blind spots	▪ Competitive intelligence synthesis ▪ Personalized emerging risk briefings ▪ Simulated scenario testing
IT & Software	▪ Boost uptime, ▪ Accelerate delivery Reduce risk	▪ Intelligent operations assistance ▪ Rapid code generation, ▪ Vulnerability identification
Marketing & Sales	▪ Boost campaign performance ▪ Accelerate deals ▪ Strengthen retention	▪ AI-powered content creation ▪ Propensity modeling ▪ Intelligent sales assistants
Customer Service	▪ Boost satisfaction, revenue, retention	▪ Intelligent virtual agents ▪ Conversational analytics ▪ Hyper-personalized marketing
Supply Chain	▪ Enhance resilience, visibility, coordination	▪ Intelligent negotiation ▪ Dynamic inventory optimization ▪ Immersive control towers
HR	▪ Boost engagement, ▪ Accelerate innovation ▪ Improve efficiency	▪ Virtual assistants ▪ Internal talent marketplaces ▪ Sentiment analysis
Finance	▪ Accelerate period-end closes ▪ Strengthen controls ▪ Uncover opportunities	▪ Transaction pattern detection ▪ Intelligent process automation ▪ Virtual assistants
Legal & Compliance	▪ Accelerate drafting ▪ Improve traceability ▪ Increase oversight	▪ Intelligent contract negotiation ▪ Augmented data mapping ▪ Personalized regulatory guidance

Table 4-2: The High-Value Business Cases Across Organizational Functions

Realizing the Business Value of Generative AI for Executives and Boards

For executives and board members, the integration of generative AI into key leadership functions such as strategy development, risk management, and governance holds substantial promise for enhancing decision-making capabilities, foresight, and operational efficiency. As top-level leaders and decision-makers consider the strategic adoption of this emerging technology, a thorough evaluation of its potential to add value in various applications is crucial. An extensive analysis of a wide range of applications has identified three distinct levels of impact:

High-Impact Applications: Certain generative AI solutions emerge as particularly influential in streamlining executive decision-making and enhancing board effectiveness. These include accelerating strategic planning cycles, bolstering risk controls, and illuminating previously unseen business opportunities. The application of AI in synthesizing competitive intelligence can dramatically revitalize strategy setting and peer benchmarking. Additionally, AI-driven personalized briefings on emerging risks and macroeconomic trends offer invaluable foresight, aiding leaders in anticipating and mitigating potential challenges. The ability of generative AI to facilitate simulated scenario testing is another area where leaders can explore various future possibilities, effectively stress-testing strategic plans in diverse scenarios.

Moderate-Impact Use Cases: Generative AI also offers significant benefits in more incremental, yet critical areas including automating stakeholder sentiment monitoring, streamlining board reporting processes, and identifying organizational weaknesses. The rapid document creation capabilities of AI can also significantly simplify the crafting of shareholder communications and executive narratives, thereby enhancing transparency and stakeholder engagement.

Emerging Applications with Cautious Potential: While promising, certain emerging applications of generative AI in autonomous strategic adjustments, unsupervised resource allocation, and comprehensive governance automation may deliver limited immediate impact. The complexity and novelty of these applications pose challenges in demonstrating short-term standalone value, but they hold potential as part of a broader digital transformation strategy.

Realizing the Business Value of Generative AI in IT and Software

Across critical technology functions like infrastructure management, software development, and security, generative AI shows immense promise to enhance productivity and efficiency. As CIOs and technology leaders explore investments in this emerging capability, evaluating the potential business value across key use cases is imperative. Based on an assessment of numerous applications, three tiers of impact have emerged:

High-Potential IT Applications: Several generative AI solutions stand out for their potential to boost uptime, accelerate delivery, and reduce risk. Intelligent operations assistance and alert correlation promise to increase infrastructure reliability at lower administrative costs. Meanwhile, rapid code generation and intelligent testing optimization have significant upside for development velocity and quality.

Vulnerability identification and access governance automation additionally help strengthen security postures.

Moderate-Potential Use Cases: More incremental yet still important potential exists in areas like virtual support agents, workflow documentation, and policy recommendation. Generative AI's rapid content creation could also simplify regulatory reporting and internal communications. Self-service dashboard creation, task prioritization, and alert routing additionally warrant consideration.

Lower-Potential Applications: While intriguing, emerging use cases around immersive system simulations, automated space design, and generalized "digital twins" may underwhelm in their near-term impact. Proving standalone value for these innovative but narrow functions could be challenging, although they may complement broader digital transformation.

Realizing the Business Value of Generative AI in Marketing and Sales

In the dynamic domains of marketing and sales, generative AI presents an opportunity to significantly elevate key functions such as campaign creation, lead management, and client engagement. For CMOs and sales leaders considering investments in this burgeoning technology, a strategic evaluation of its potential business impact across various applications is essential. An extensive analysis of numerous applications reveals a tiered structure of impact:

High-Impact Marketing & Sales Applications: Certain generative AI solutions have emerged as particularly potent, with the capability to enhance campaign effectiveness, speed up deal progression, and bolster customer retention. AI-driven content generation holds the promise of creating more engaging, cost-efficient promotional materials. Propensity modeling and hyper-personalization strategies stand out for their ability to effectively convert new opportunities. Additionally, intelligent virtual sales assistants offer invaluable support to sales representatives, enabling deeper and more consistent client interactions.

Moderate-Impact Use Cases: In areas such as personalized bundle recommendations, automated lead scoring, and win/loss analysis, generative AI offers significant, albeit more incremental, benefits. Rapid reporting and automated document creation capabilities of AI can substantially reduce time spent on non-strategic tasks, thereby increasing time for critical selling activities. Predictive analytics in identifying churn risks and recommending next-best actions also play an important role in enhancing sales strategies.

Emerging Applications with Lower Immediate Impact: Emerging applications like fully autonomous lead prospecting, unsupervised bid approvals, and broad-scale campaign orchestration, while innovative, may initially deliver limited impact. These complex functions present challenges in demonstrating immediate standalone value but could be instrumental in complementing a broader technological transformation in sales and marketing.

Realizing the Business Value of Generative AI in Customer Services

Across critical functions like service delivery, customer analytics, and journey orchestration, generative AI shows immense promise to enhance experiences, loyalty, and efficiency. As Chief Customer Officers and CX leaders explore investments in this emerging capability, evaluating the potential business value across key use cases is imperative. Based on an assessment of numerous applications, three tiers of impact have emerged:

High-Potential Applications: Several generative AI solutions stand out for their potential to boost satisfaction, revenue, and retention. Intelligent virtual agents promise to resolve inquiries faster while conversational analytics uncover product and journey improvements. Meanwhile, hyper-personalized marketing and seamless post-purchase care unlock growth through relevance and trust. Sentiment tracking and predictive modeling additionally give brands an unprecedented understanding of evolving customer needs.

Moderate-Potential Use Cases: More incremental yet still important potential exists in areas like automated order modifications, personalized bundle recommendations, and dynamic creative optimizations. Generative AI's rapid content creation could also simplify regulatory disclosures and improve self-service resources. Workflow assistance for agents and marketers additionally warrants consideration.

Lower-Potential Applications: While intriguing, emerging use cases around fully autonomous customer care, product engineering changes based solely on feedback, and broad customer experience oversight may underwhelm in their near-term impact. Proving standalone value for these innovative but highly complex functions could be challenging, although they may complement broader digital transformation.

Realizing the Business Value of Generative AI in Sourcing, Procurement, and Supply Chain

Across critical functions like supplier management, inventory planning, and logistics, generative AI shows immense promise to enhance resilience, visibility, and coordination. As Chief Procurement Officers and supply chain leaders explore investments in this emerging capability, evaluating the potential business value across key use cases is imperative. Based on an assessment of numerous applications, three tiers of impact have emerged:

High-Potential Applications: Several generative AI solutions stand out for their potential to mitigate risk, accelerate innovation, and radicalize efficiency. Intelligent negotiation promises to optimize procurement costs and contracts through data synthesis and strategy automation. Dynamic inventory optimization and demand forecasting also hold tremendous value. Meanwhile, immersive control towers offer end-to-end visibility for proactive disruption response.

Moderate-Potential Use Cases: More incremental yet still important potential exists in supplier discovery, virtual buying assistants, and customer service chatbots. Conversation AI could also simplify inquiries and transactions. Spend analysis, contract extraction, and quality assurance enhancements additionally warrant consideration.

Lower-Potential Applications: While intriguing, emerging use cases around completely lights-out production, generalized "digital twin" models, and autonomous last-mile delivery may underwhelm in their near-term impact. Proving standalone value for these innovative but narrow functions could be challenging, although they may complement broader digital transformation.

Realizing the Business Value of Generative AI in HR

Across core talent functions like hiring, development, and retention, generative AI shows immense promise to enhance productivity and transformation. As HR leaders explore investments in this emerging capability, evaluating the potential business value across key use cases is critical. Based on an assessment of numerous applications, three tiers of impact have emerged:

High-Potential HR Applications: Several generative AI solutions stand out for their potential to boost engagement, accelerate innovation, and radicalize efficiency. Virtual assistants promise to scale student support and engagement at lower administrative costs. Meanwhile, optimized internal talent marketplaces and personalized learning matches have significant upside for agility and capability building. Sentiment analysis and voice of the employee capabilities additionally enable institutions to amplify their societal impact.

Moderate-Potential Use Cases: More incremental yet still important potential exists in areas like candidate experience enhancement, automated job description creation, and real-time HR query handling. Generative AI's rapid content creation could also simplify regulatory reporting and employee communications. Recommendation algorithms around flight risk intervention, assessments, and nudging additionally warrant consideration.

Lower-Potential Applications: While intriguing, emerging use cases virtual coaching, interview summarization, and performance review automation may underwhelm in their near-term potential. Proving value for these innovative but narrow functions could be challenging, although they may complement broader digital transformation.

Realizing the Business Value of Generative AI in Finance

Across critical finance functions like reporting, transaction processing, and data analytics, generative AI shows immense promise to enhance productivity, insights, and oversight. As CFOs and finance leaders explore investments in this emerging capability, evaluating the potential business value across key use cases is imperative. Based on an assessment of numerous applications, three tiers of impact have emerged:

High-Potential Finance Applications: Several generative AI solutions stand out for their potential to accelerate period-end closes, strengthen internal controls, and uncover growth opportunities. Transaction pattern detection and anomaly identification promise to reduce fraud and errors. Meanwhile, intelligent process automation and augmented analytics have tremendous value for unlocking capacity and elevating planning. Virtual assistants additionally give finance teams ubiquitous, on-demand support.

Moderate-Potential Use Cases: More incremental yet still important potential exists in areas like regulatory document analysis, external spend optimization, and predictive modeling. Generative AI's rapid content creation could also simplify narrative financial reporting and management communications. Process mining, balance sheet forecasting, and auditing assistances additionally warrant consideration.

Lower-Potential Applications: While intriguing, emerging use cases around fully autonomous accounting operations, self-optimizing hedging strategies, and broad finance workflow orchestration may underwhelm in their near-term impact. Proving standalone value for these innovative but highly complex functions could be challenging, although they may complement broader digital transformation.

Realizing the Business Value of Generative AI in Legal, Privacy, and Compliance

Across critical functions like contract management, data governance, and regulatory adherence, generative AI shows immense promise to enhance efficiency, transparency, and risk mitigation. As Chief Legal Officers and compliance leaders explore investments in this emerging capability, evaluating the potential business value across key use cases is imperative. Based on an assessment of numerous applications, three tiers of impact have emerged:

High-Potential Legal Applications: Several generative AI solutions stand out for their potential to accelerate contract drafting, improve data traceability, and increase oversight rigor. Intelligent negotiation promises to optimize terms and conditions through historical data synthesis. Meanwhile, augmented data mapping and audit trailing offer end-to-end visibility into information flows and usage. Personalized regulatory guidance and reporting additionally help strengthen privacy postures and governance.

Moderate-Potential Use Cases: More incremental yet still important potential exists in areas like virtual legal assistants, litigation brief assembly, and IP portfolio optimization. Generative AI's rapid content creation could also simplify policy documentation and statement generation. Spend analysis, matter prioritization, and risk detection enhancements additionally warrant consideration.

Lower-Potential Applications: While intriguing, emerging use cases around fully automated contract execution, unsupervised data deletion triggers, and broad operational compliance may underwhelm in their near-term impact. Proving standalone value for these innovative but highly complex functions could be challenging, although they may complement broader digital transformation.

Evaluating the Business Value Holistically

Recent studies indicate that a staggering 80% of AI initiatives fall short in delivering tangible business value, primarily due to a narrow focus on the technological aspects rather than their alignment with strategic growth objectives. As generative AI begins to permeate various business sectors, it becomes imperative for organizations to adopt a holistic strategy in their investment planning and governance of this technology. It's not sufficient to merely assess the value potential of AI within isolated business functions; the broader, systemic impacts must also be considered, especially given the pervasive nature of this emerging general-purpose technology.

The application of generative AI, for instance, might boost a sales team's performance through more targeted outreach efforts. However, if implemented without careful consideration, the same technology could inadvertently harm brand integrity or customer trust. To fully leverage the diverse benefits of generative AI, a comprehensive, cross-functional collaboration is essential from the very beginning. Organizations must cultivate a shared understanding of the potential risks and collectively define clear objectives. Establishing robust principles for AI model development and maintaining rigorous human oversight are crucial steps in harnessing the extensive promise of generative AI.

This holistic approach ensures that while individual functions reap the benefits of AI, the organization as a whole moves forward cohesively, aligning technological advancements with its core values and strategic goals.

The AI Business Value Octagon: A Comprehensive framework

In an era marked by rapid technological advancements and shifting market dynamics, businesses are increasingly turning to generative AI (GenAI) to drive innovation and growth. To harness the full potential of GenAI, a nuanced and multifaceted approach is essential. The "**AI Business Value Octagon**" is a pioneering framework developed through a deep analysis of several hundred business cases across a wide array of industries and organizational functions. These cases, meticulously detailed in the preceding sections of this book, provide the empirical foundation for this comprehensive framework.

Each dimension of the Octagon Framework represents a critical area where GenAI can significantly influence business outcomes. These dimensions are Productivity Gains, Innovative Capabilities, Total Experience Enhancement, Growth Potential, Cost Reduction, Risk Mitigation, Competitive Differentiation, and Sustainability and Compliance. The framework intricately weaves these dimensions together, offering a holistic view of how GenAI can be strategically integrated into various aspects of business operations.

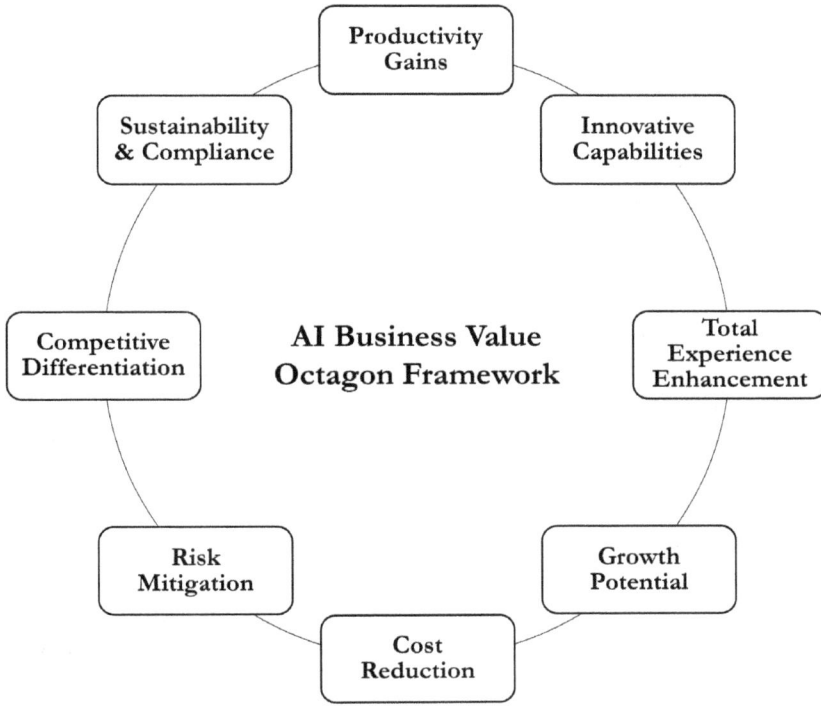

Figure 4-1: The AI Business Value Octagon with Eight Value Dimensions

1. Productivity Gains: Harnessing GenAI for Operational Excellence

Within the framework of the AI Business Value Octagon, the dimension of 'Productivity Gains' holds a central role. It encapsulates the transformative ability of generative AI (GenAI) to elevate the efficiency and effectiveness of various business operations, marking a new era in operational management.

Core Aspects of Productivity Gains through GenAI

1. **Automation of Routine Tasks**: GenAI's prowess in automating mundane, time-intensive tasks is unparalleled. This capability not only liberates human resources from repetitive duties but also reallocates their talents to more complex, thought-driven tasks. This shift allows for a higher level of creativity and strategic thinking within the workforce. Practical applications of this facet include the automation of report generation, data entry processes, and handling standard customer service queries.

2. **Enhancement of Output Quality**: GenAI's impact extends beyond process acceleration to significantly uplift the quality of outputs. Leveraging advanced algorithms, GenAI can outperform human accuracy and consistency, substantially reducing errors and enhancing the caliber of end products. In industries like manufacturing, this advantage manifests as a decrease in product defects and a boost in overall reliability and consumer trust.

3. **Optimization of Resource Utilization**: GenAI excels in analyzing trends and forecasting requirements, enabling smarter, more efficient resource allocation. This facet covers improved inventory management, optimized staffing schedules, and the strategic distribution of budgets and materials. The outcome is a more cost-effective, efficient operation with minimized waste and maximized output.

4. **Efficient Scalability of Operations**: A critical advantage of GenAI is its ability to facilitate business scalability without necessitating a corresponding increase in expenses or resources. This attribute is particularly invaluable for businesses in growth phases or those encountering variable demand, offering a pathway to expansion that is both sustainable and manageable.

5. **Minimization of Manual Effort and Error**: GenAI significantly reduces the need for manual intervention in complex tasks, consequently diminishing the likelihood of human error. This reduction has a ripple effect – it not only streamlines processes but also boosts customer satisfaction and lowers the costs associated with error correction and quality control.

Each aspect of productivity gains is supported by real-world examples and case studies from the extensive research conducted. For instance, a case study from the retail sector might demonstrate how GenAI-driven predictive analytics has streamlined inventory management, leading to reduced waste and improved profit margins. Similarly, in the healthcare sector, the use of GenAI for patient data management and diagnostics has not only saved time but also improved patient outcomes.

For businesses to fully harness these benefits, a strategic approach to GenAI adoption is imperative. This involves identifying areas with the highest impact potential, seamlessly integrating GenAI solutions with existing systems, encouraging continuous learning and adaptation of these systems, and ensuring that employees are well-equipped to collaborate with these new technological tools.

The 'Productivity Gains' dimension of the AI Business Value Octagon highlights the transformative impact of GenAI on business operations. By automating routine tasks, enhancing output quality, optimizing resource allocation, efficiently scaling operations, and reducing manual effort and error, GenAI stands as a powerful tool for operational excellence. The adoption and strategic implementation of GenAI, backed by real-world examples and a deep understanding of its capabilities, can lead businesses to new heights of productivity and efficiency.

2. Innovative Capabilities: Driving Business Transformation with GenAI

In the AI Business Value Octagon, 'Innovative Capabilities' stands as a crucial dimension. This section delves into how generative AI (GenAI) fosters innovation, not just as a tool for enhancing existing processes, but as a catalyst for creating new business paradigms and disrupting traditional market practices.

Facets of Innovation Enabled by GenAI

1. **Creation of Novel Products and Services**: GenAI empowers businesses to develop innovative products and services that were previously unimaginable. This includes leveraging AI for personalized product design, creating intelligent service offerings, or developing entirely new customer experiences. These innovations can open up untapped markets and create new revenue streams.

2. **Disrupting Market Norms**: GenAI has the potential to redefine industry standards and disrupt existing market norms. By introducing cutting-edge solutions that challenge traditional business models, companies can gain a significant competitive edge. This disruption is not limited to technological advancements but extends to changing consumer behaviors and expectations.

3. **Facilitating New Business Models**: GenAI enables organizations to explore and establish new business models. This could involve AI-driven subscription services, dynamic pricing models, or new forms of customer engagement and interaction. These models can lead to more sustainable and scalable business practices.

4. **Enhancing Decision-Making Processes**: With its ability to analyze vast amounts of data and generate insights, GenAI becomes an invaluable asset for strategic decision-making. It provides business leaders with the tools to make more informed, data-driven decisions, paving the way for innovative strategies and growth opportunities.

5. **Accelerating Research and Development (R&D)**: GenAI significantly speeds up the R&D process, allowing for quicker innovation cycles. This acceleration is critical in industries where staying ahead of technological trends is vital. AI can identify patterns and possibilities that human researchers may overlook, leading to groundbreaking discoveries.

Each aspect of the 'Innovative Capabilities' dimension is underpinned by real-world examples and case studies. For instance, a tech company might use GenAI to develop a revolutionary user interface that adapts to individual user behaviors, setting a new standard in personal technology. In the pharmaceutical industry, GenAI could be instrumental in discovering new drugs, drastically reducing the time and cost associated with bringing them to market.

To strategically implement GenAI for nurturing innovation, organizations must cultivate an environment that champions creativity and experimentation. This involves investing in the necessary skills and training, particularly in areas of data science and AI ethics, to empower teams to effectively leverage GenAI. Encouraging cross-functional collaboration is also vital, as innovation often emerges from the synergy of diverse perspectives and expertise. Additionally, staying abreast of the latest developments in AI and related technologies is crucial, as this knowledge can spark new ideas and inspire groundbreaking applications. This comprehensive approach ensures that GenAI is not only integrated into the fabric of the organization but also serves as a dynamic catalyst for continuous innovation and growth.

The 'Innovative Capabilities' segment of the AI Business Value Octagon underscores the transformative power of GenAI in driving business innovation. By enabling the creation of new products and services, disrupting market norms, facilitating novel business models, enhancing decision-making, and accelerating R&D, GenAI positions itself as an essential driver of future business success. Embracing these capabilities with a strategic, informed, and collaborative approach can lead organizations towards a future rich with innovation and growth opportunities.

3. Total Experience Enhancement: Revolutionizing Stakeholder Interactions with GenAI

In the AI Business Value Octagon, the 'Total Experience Enhancement' dimension plays a pivotal role. It encapsulates how generative AI (GenAI) can be leveraged to holistically enhance the experiences of all stakeholders — employees, customers, and partners alike. This approach is not just about improving individual touchpoints but about transforming the entire journey of interactions they have with the organization.

Key Elements of Total Experience Enhancement through GenAI

1. **Empowering Employees for Greater Productivity and Satisfaction**: GenAI can significantly elevate the employee experience by automating mundane tasks, providing intelligent support systems, and offering personalized learning and development opportunities. This leads to increased job satisfaction, higher levels of engagement, and a more innovative and productive workforce.

2. **Enriching Customer Interactions**: In the realm of customer experience, GenAI introduces a new level of personalization and responsiveness. From AI-driven recommendation systems to intelligent customer support bots, GenAI can create more intuitive, engaging, and satisfying customer journeys. This enhances customer loyalty and can lead to a better brand perception.

3. **Strengthening Partner and Stakeholder Relationships**: GenAI can also be instrumental in improving interactions with business partners and other stakeholders. By providing more accurate and timely data, automating collaborative processes, and facilitating better communication, GenAI strengthens these relationships, leading to more successful and sustainable partnerships.

4. **Integrating Experiences Across Channels**: A key advantage of GenAI is its ability to integrate experiences across various channels and touchpoints. This seamless integration ensures a consistent and cohesive experience, whether stakeholders are interacting online, in-person, or through other mediums.

5. **Predictive and Proactive Engagement**: GenAI's ability to analyze data and predict trends allows organizations to proactively address stakeholder needs and preferences. This can transform the way organizations engage with their stakeholders, moving from a reactive to a proactive stance, and delivering experiences that are not only satisfying but also delightfully unexpected.

Real-world examples underpin each element of the 'Total Experience Enhancement' dimension. For instance, a retail company might use GenAI to offer personalized shopping experiences to customers, resulting in increased sales and customer loyalty. In a corporate setting, GenAI-driven tools could be used to streamline workflow and improve communication, leading to a more engaged and efficient workforce.

The 'Total Experience Enhancement' dimension of the AI Business Value Octagon highlights the profound impact of GenAI in transforming the experiences of all stakeholders. By enriching employee, customer, and partner interactions, integrating experiences across channels, and enabling predictive and proactive engagement, GenAI emerges as a vital tool for businesses striving to create a more connected, responsive, and satisfying ecosystem. This holistic enhancement of experiences not only improves individual satisfaction but also contributes significantly to the overall health and success of the organization.

4. Growth Potential: Catalyzing Revenue Generation with GenAI

In the AI Business Value Octagon, 'Growth Potential' is a crucial dimension, emphasizing how generative AI (GenAI) can significantly drive business growth and increase revenue generation. This section delves into the diverse ways in which GenAI not only propels expansion but also directly contributes to enhancing revenue streams.

Key Drivers of Growth and Revenue through GenAI

1. **Market Expansion and Diversification**: GenAI is instrumental in identifying and penetrating new markets, often unlocking revenue streams in previously untapped areas. By leveraging data-driven insights to understand new customer segments and preferences, businesses can diversify their offerings, reducing market-specific risks and boosting revenue potential.

2. **Enhancing Customer Base and Loyalty**: The advanced personalization capabilities of GenAI play a key role in attracting and retaining customers, directly influencing revenue growth. Tailored experiences, predictive services, and enhanced engagement strategies not only solidify customer loyalty but also attract new customers, thus expanding the revenue base.

3. **Streamlining Operations for Scalability and Efficiency**: GenAI contributes to scalable growth by automating and optimizing operations, leading to cost savings and improved efficiency. This operational leverage allows businesses to expand rapidly, enhancing their revenue potential without a corresponding increase in operational costs.

4. **Driving Innovation for Competitive Edge**: Innovation, fueled by GenAI, opens new avenues for revenue generation. By developing new products, services, and business models, companies can tap into unex-

plored markets and create additional sources of income, maintaining a competitive edge in the market.

5. **Data-Driven Strategies for Revenue Optimization**: GenAI's prowess in analyzing vast datasets enables businesses to make strategic decisions that are crucial for revenue growth. This includes identifying lucrative investment opportunities, optimizing pricing strategies, and uncovering new sales channels.

This dimension's impact is illustrated through various real-world examples and case studies, demonstrating how businesses have successfully harnessed GenAI for revenue growth. For instance, a case study may showcase how a retail company used GenAI for dynamic pricing, leading to increased sales and profit margins. Another example could highlight how a B2B company leveraged GenAI for lead generation and customer segmentation, resulting in higher conversion rates and revenue.

The 'Growth Potential' dimension of the AI Business Value Octagon highlights the critical role of GenAI in driving business growth and increasing revenue. By enabling market expansion, customer base enhancement, operational scalability, continuous innovation, and strategic decision-making, GenAI emerges as a vital tool for businesses aiming to boost their revenue and secure a dominant position in their respective markets. The strategic implementation of GenAI, informed by empirical evidence and success stories, paves the way for organizations to not only grow but also significantly enhance their revenue generation capabilities in an evolving business landscape.

5. Cost Reduction: Streamlining Expenses with GenAI

In the AI Business Value Octagon, the 'Cost Reduction' dimension emphasizes how generative AI (GenAI) can be strategically utilized to streamline costs and enhance financial efficiency. This section explores how GenAI can lead to significant savings by optimizing various aspects of business operations, contributing directly to the bottom line.

Strategic Approaches to Cost Reduction through GenAI

1. **Efficiency in Operations**: GenAI plays a crucial role in automating routine and complex tasks, leading to a more efficient use of resources. This automation reduces the need for manual labor in certain areas, consequently lowering labor costs. Additionally, GenAI-driven process optimization minimizes operational inefficiencies, further reducing opera-

tional expenses.

2. **Reduced Error Rates and Waste**: GenAI's advanced analytics and predictive capabilities lead to a significant reduction in errors and waste. In manufacturing, for instance, GenAI can optimize production lines to minimize material wastage. In service industries, error reduction means fewer resources spent on rectifying mistakes, leading to cost savings.

3. **Energy and Resource Optimization**: By analyzing usage patterns and predicting future needs, GenAI can optimize the consumption of energy and other resources. This not only contributes to cost savings but also aligns with sustainable business practices, potentially reducing environmental impact and associated costs.

4. **Optimized Supply Chain Management**: GenAI enhances supply chain efficiency by forecasting demand, optimizing inventory levels, and streamlining logistics. This results in reduced storage costs, minimized overstocking or stockouts, and more efficient distribution, all of which contribute to lowered operational costs.

5. **Strategic Decision Making for Cost Management**: With its ability to process and analyze vast amounts of data, GenAI aids in making strategic decisions that can lead to cost savings. This includes identifying areas of excessive spending, optimizing budget allocations, and making informed investment decisions.

Real-world examples and case studies underline the 'Cost Reduction' dimension. For instance, a retail company might employ GenAI for inventory management, significantly reducing costs associated with overstocking and stock obsolescence. Another example could be a logistics company using GenAI to optimize route planning, leading to fuel savings and reduced wear-and-tear on vehicles.

The 'Cost Reduction' dimension of the AI Business Value Octagon underscores the significant impact GenAI can have in reducing operational and production costs. By streamlining processes, reducing errors, optimizing resource usage, enhancing supply chain management, and aiding strategic financial decisions, GenAI proves to be an invaluable asset for cost-efficient business operations. This dimension not only contributes to immediate cost savings but also fosters a culture of efficiency and sustainability, driving long-term financial health for businesses in a competitive global marketplace.

6. Risk Mitigation: Enhancing Security and Stability with GenAI

In the AI Business Value Octagon, the 'Risk Mitigation' dimension is vital, highlighting how generative AI (GenAI) can significantly bolster an organization's ability to identify, assess, and mitigate various forms of risk. This section delves into the multifaceted role of GenAI in enhancing risk management practices, thereby ensuring business stability and security.

Facets of Risk Mitigation Through GenAI

1. **Predictive Risk Analysis**: GenAI excels in predicting potential risks by analyzing vast datasets and identifying patterns that might elude human analysis. This capability is crucial in sectors like finance, where predictive models can foresee market fluctuations, or in manufacturing, where it can anticipate equipment failures, thus preemptively addressing issues before they escalate.

2. **Enhanced Cybersecurity Measures**: As cyber threats become more sophisticated, GenAI provides advanced tools for cybersecurity. By continuously learning from new data, GenAI models can detect, analyze, and respond to cyber threats more efficiently, staying ahead of hackers and reducing the risk of data breaches and other cyber attacks.

3. **Improved Compliance and Regulatory Adherence**: Navigating the complex landscape of legal and regulatory compliance is a significant challenge for businesses. GenAI can automate and enhance the monitoring of compliance requirements, ensuring that the organization adheres to relevant laws and regulations, thereby mitigating legal risks.

4. **Operational Risk Management**: GenAI aids in identifying and addressing operational risks, such as supply chain disruptions or process inefficiencies. By providing real-time insights and forecasts, it enables businesses to make informed decisions, ensuring operational continuity and resilience.

5. **Strategic Decision-Making Support**: In strategic planning and decision-making, GenAI contributes to risk mitigation by offering data-driven insights. This support helps in assessing the viability of business strategies, evaluating potential threats, and making informed decisions that minimize risk exposure.

This dimension is supported by real-world examples and case studies demonstrating GenAI's impact on risk mitigation. For instance, a financial institution using GenAI for fraud detection and prevention has significantly reduced fraudulent activities, protecting both the institution and its customers. Similarly, a case study in the healthcare sector might illustrate how GenAI has improved patient safety by predicting and preventing medical errors.

The 'Risk Mitigation' dimension of the AI Business Value Octagon emphasizes the critical role of GenAI in safeguarding against various risks. By enhancing predictive risk analysis, bolstering cybersecurity, ensuring regulatory compliance, managing operational risks, and supporting strategic decisions, GenAI emerges as an essential component of modern risk management strategies. Its implementation helps organizations not only in averting potential crises but also in fostering a more secure, stable, and resilient business environment in an increasingly uncertain world.

7. Competitive Differentiation: Establishing a Unique Market Position with GenAI

Within the AI Business Value Octagon, the 'Competitive Differentiation' dimension highlights the crucial role of generative AI (GenAI) in helping businesses establish a distinct and advantageous position in the marketplace. This section explores how GenAI can be leveraged to create unique value propositions, differentiate products and services, and ultimately gain a competitive edge.

Core Strategies for Competitive Differentiation through GenAI

1. **Innovative Product and Service Offerings**: GenAI enables the creation of novel and innovative products and services, setting businesses apart from their competitors. By harnessing AI's power, companies can introduce features and functionalities that are not only cutting-edge but also highly tailored to customer needs, fostering a unique market identity.

2. **Personalized Customer Experiences**: GenAI's capability to analyze customer data and predict preferences allows businesses to offer highly personalized experiences. This level of personalization can significantly differentiate a brand in a crowded market, leading to enhanced customer loyalty and brand advocacy.

3. **Enhanced Quality and Performance**: GenAI can significantly improve the quality and performance of products and services. In sectors like manufacturing and software, AI-driven quality control and performance op-

timization can result in superior offerings, distinguishing a company from its competitors.

4. **Data-Driven Marketing and Sales Strategies**: GenAI's advanced analytics enable more effective and targeted marketing and sales strategies. By understanding customer behaviors and market trends at a granular level, businesses can craft campaigns and sales approaches that resonate more effectively with their target audience, thereby gaining a competitive advantage.

5. **Operational Excellence and Efficiency**: By streamlining operations and enhancing efficiency, GenAI contributes to a more agile and responsive business model. This operational excellence not only reduces costs but also allows businesses to respond swiftly to market changes, further differentiating them in the eyes of customers and partners.

The impact of the 'Competitive Differentiation' dimension is illustrated with real-world examples and case studies. For example, a tech company might use GenAI to develop an AI-powered personal assistant, far surpassing the capabilities of existing products in the market. Another instance could be a retail brand using GenAI for inventory and supply chain optimization, significantly reducing delivery times and improving customer satisfaction.

The 'Competitive Differentiation' dimension of the AI Business Value Octagon emphasizes the transformative impact of GenAI in securing a competitive advantage. Through innovative offerings, personalized experiences, superior quality, data-driven strategies, and operational excellence, GenAI empowers businesses to stand out in their respective industries. Embracing GenAI strategically, as highlighted by empirical evidence and success stories, paves the way for organizations to not just compete but lead in an increasingly dynamic and competitive global marketplace.

8. Sustainability and Compliance: Aligning GenAI with Ethical and Environmental Standards

In the AI Business Value Octagon, 'Sustainability and Compliance' is a critical dimension that emphasizes the importance of aligning generative AI (GenAI) with ethical norms, environmental sustainability, and regulatory compliance. This section discusses how GenAI can be integrated into business practices in a way that upholds and advances these crucial standards.

Integrating GenAI with Sustainability and Compliance Objectives

1. **Ethical AI Deployment**: Ethical considerations are paramount in the deployment of GenAI. This involves ensuring that AI systems are transparent, fair, and unbiased, and that they respect user privacy and data security. Ethical AI deployment not only aligns with moral imperatives but also builds trust with customers and other stakeholders.

2. **Environmental Sustainability through AI**: GenAI can play a significant role in promoting environmental sustainability. By optimizing resource use, reducing waste, and enhancing energy efficiency, AI systems can help reduce the environmental footprint of businesses. This includes everything from streamlining logistics to minimize carbon emissions to employing AI in the development of sustainable materials and processes.

3. **Regulatory Compliance and Risk Management**: In an era of increasing regulatory scrutiny, GenAI can assist businesses in maintaining compliance with various legal and industry-specific standards. AI systems can be used to monitor and report on compliance-related data, thus reducing the risk of regulatory breaches and the associated financial and reputational costs.

4. **Enhancing Transparency and Accountability**: Transparency and accountability in AI operations are essential for compliance and building stakeholder confidence. GenAI can help in documenting decision-making processes and providing clear audit trails, which are crucial for demonstrating compliance and ethical business practices.

5. **Social Responsibility and Community Engagement**: GenAI can also contribute to a company's social responsibility initiatives, such as by aiding in the analysis and implementation of community engagement strategies or in addressing social issues through innovative AI applications.

The 'Sustainability and Compliance' dimension is supported by real-world examples where GenAI has been successfully aligned with these goals. For instance, a case study might illustrate how a manufacturing company used GenAI to reduce its carbon footprint significantly, or how a financial institution leveraged AI to ensure comprehensive compliance with global financial regulations.

The 'Sustainability and Compliance' dimension of the AI Business Value Octagon underscores the importance of responsibly integrating GenAI within business operations. By focusing on ethical AI deployment, environmental sustainability, regulatory compliance, transparency, and social responsibility, businesses can ensure that their use of GenAI not only drives performance and profitability but also aligns with broader societal values and standards. This approach to GenAI adoption is essential for building a sustainable, compliant, and ethically sound business future.

Effectively Employing the Octagon Framework

This essential section of the book is specifically tailored for business leaders, offering a detailed strategic roadmap for the effective implementation and utilization of the AI Business Value Octagon framework. The guide is designed to help leaders navigate the complexities of integrating GenAI into their organizations, ensuring alignment with broader corporate objectives and maximizing the potential benefits across all Octagon dimensions.

Strategies for Prioritizing Investments

1. **Alignment with Corporate Goals**: Leaders are provided with methodologies to align GenAI initiatives with their organization's overarching goals. This involves evaluating how each dimension of the Octagon framework can support strategic objectives, be it growth, innovation, cost reduction, or any other key area.

2. **Resource Allocation**: The guide offers insights on judiciously allocating resources - including financial, human, and technological - to GenAI projects. This includes identifying areas where investment in GenAI can yield the highest returns and understanding how to balance short-term benefits with long-term strategic gains.

Metrics for Measuring Impact

1. **Developing KPIs and Benchmarks**: For each dimension of the Octagon framework, a comprehensive set of Key Performance Indicators (KPIs) and benchmarks is provided. These metrics enable leaders to measure the impact of GenAI initiatives quantitatively, assessing everything from productivity gains and cost savings to improvements in customer and employee satisfaction.

2. **Continuous Monitoring and Adjustment**: The guide emphasizes the importance of continuous performance tracking and the need to adapt strategies based on these metrics, ensuring that GenAI initiatives remain aligned with evolving business goals and market conditions.

Communicating Value to Stakeholders

1. **Effective Communication Strategies**: Leaders are equipped with strategies to effectively communicate the value and potential of GenAI to various stakeholders. This includes crafting clear messages that articulate the benefits of GenAI initiatives, tailored to resonate with different groups from operational teams to the C-suite.

2. **Building Buy-In**: The guide includes techniques for building buy-in and fostering a shared vision for GenAI across the organization, recognizing that stakeholder engagement is crucial for the successful adoption and implementation of GenAI technologies.

Fostering Holistic Development

1. **Balancing Financial, Social, and Environmental Goals**: Insights are provided on how to ensure that GenAI deployments contribute positively not just to financial performance but also to employee welfare and environmental sustainability, fostering a holistic approach to business development.

2. **Ethical Considerations and Sustainability**: The guide stresses the importance of ethical considerations in GenAI deployment, along with strategies to ensure that GenAI applications are sustainable and aligned with the organization's corporate social responsibility goals.

This guide serves as a comprehensive manual for leaders to strategically harness the power of GenAI, guiding them through the process of aligning investments with business objectives, measuring impact through tailored metrics, effectively communicating the value of GenAI, and ensuring holistic development that balances financial success with ethical and sustainable practices.

Prioritizing AI Value to Strategic Focus

In integrating the AI Business Value Octagon framework, a key consideration for business leaders is the prioritization of AI value dimensions in alignment with their company's strategic focus. Understanding that not all dimensions of AI value will be equally relevant to every organization, this approach involves making informed investment tradeoffs to enhance critical differentiators. Drawing on Treacy & Wiersema's Value Disciplines Model, which categorizes market positioning foci into Operational Excellence, Product Leadership, and Customer Intimacy, leaders can more effectively align AI initiatives with their primary business discipline.

- **Operational Excellence:** Companies focusing on Operational Excellence can leverage AI to enhance efficiency and flawless execution. For instance, employing generative design in warehouse layout and product flow optimization can significantly boost storage density and reduce picking times, with predictive maintenance systems minimizing equipment downtime through early failure detection.

- **Product Leadership:** For organizations prioritizing Product Leadership, AI can drive cutting-edge offerings and continuous innovation. An example is the use of generative AI in product development, where AI experiments with CAD models and consumer data to create innovative product designs, pushing the boundaries of ergonomics, style, and production feasibility.

- **Customer Intimacy:** For businesses centered on Customer Intimacy, AI can be used to offer bespoke personalization and deepen customer relationships. AI applications like a stylist chatbot that suggests daily personalized outfits based on the client's preferences, schedule, and mood, or conversational AI that analyses vocal patterns for customer satisfaction, can significantly enhance the customer experience.

In each case, AI delivers value aligned with the company's differentiation strategy, whether it be operational efficiency, product innovation, or customer centricity. Implementing these strategies requires cross-functional collaboration, where AI is applied to amplify a specific aspect of the company's existing strategic positioning.

By aligning AI initiatives with the primary market positioning focus - Operational Excellence, Product Leadership, or Customer Intimacy - leaders can sharpen capital allocation and enhance their company's competitive edge. The Octagon framework, therefore, not only guides in the comprehensive adoption of AI but also in its strategic alignment with the core business focus, ensuring that AI investments are directly contributing to the organization's overarching goals.

Conclusion

As we conclude this exploration of the AI Value Pillar and the AI Business Value Octagon framework, it's clear that the transformative power of AI in business is both immense and multifaceted. Through our journey across various industry sectors and organizational functions, we have seen how AI can be a game-changer, driving efficiency, innovation, and growth.

The AI Value Pillar, complemented by the AI Business Value Octagon, provides a comprehensive guide for businesses to strategically navigate the complex landscape of AI adoption. These frameworks serve as a beacon, directing leaders towards AI initiatives that align with their core business objectives and strategic vision. By prioritizing investments in AI based on these frameworks, organizations can close the Value Gap, ensuring that AI adoption is not just a technological upgrade but a strategic enhancement to their business model.

In conclusion, the journey towards harnessing AI's full potential is ongoing and ever-evolving. By leveraging the insights and strategic frameworks outlined in this chapter, businesses can not only anticipate and navigate the challenges of AI adoption but also capitalize on its myriad opportunities. As we look to the future, it is evident that AI will continue to be a key driver of business transformation, and those who adeptly integrate it into their strategic plans will be well-positioned to lead in their respective industries.

Chapter 5

AI Total Cost of Ownership (TCO) Analysis

"The greatest ownership of all is to glance around and understand." —
William Stafford

I n the journey towards adopting Generative AI, a common challenge faced by
many companies is the difficulty in accurately estimating the costs involved. This
uncertainty can create significant hurdles for business leaders who are tasked with
making informed budgeting and investment decisions. The complexity and novelty
of Generative AI technologies mean that the associated costs extend beyond simple
procurement or licensing fees. They encompass a range of factors including data infrastructure, engineering expertise, integration processes, and ongoing maintenance
and oversight.

This lack of clarity often leads to underestimation or misjudgment of the Total
Cost of Ownership (TCO), which can have profound implications on the strategic
and financial planning of an organization. As a result, leaders find themselves in a
challenging position, trying to balance the promising benefits of Generative AI —
such as increased productivity and customer satisfaction — with the opaque and
potentially substantial investment required.

To navigate these complexities, it is imperative for businesses to conduct a detailed
and comprehensive TCO analysis. This analysis should not only account for immediate expenditure but also consider the long-term financial commitments required
for the effective deployment and scaling of Generative AI technologies. By doing so,
companies can gain a clearer understanding of the financial implications, enabling
leaders to make more informed and strategic investment decisions.

Generative AI
TCO
Analysis

GenAI Tools &
Platform Access
Costs

Prompt
Engineering
Costs

Talent Costs

AI Regulations
Compliance
Costs

Inference Costs

Operations
Costs

Fine-Tuning
Costs

Data
Management
Costs

Infrastructure
Costs

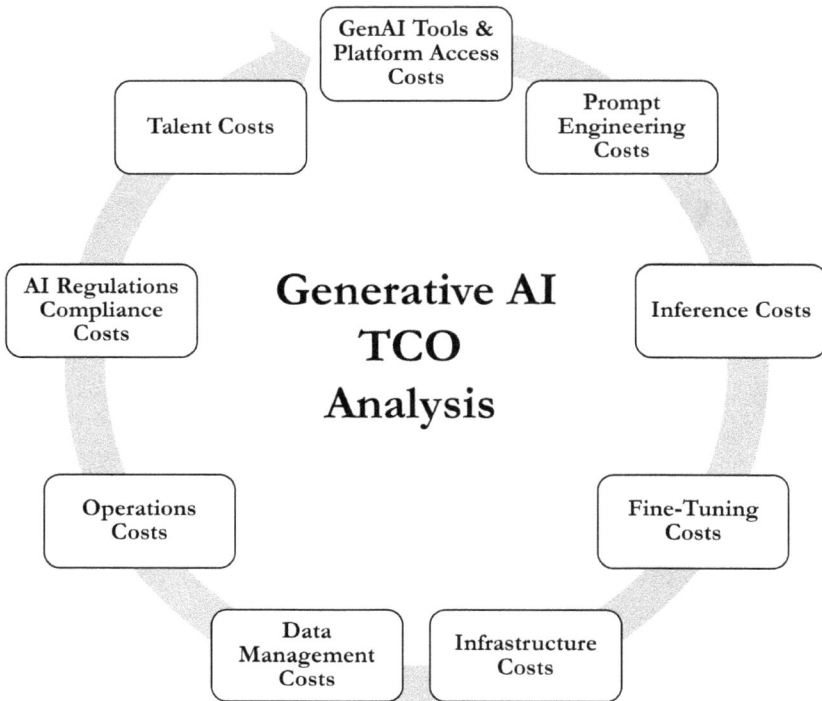

Figure 5-1: Nine Types of Costs Associated with Adopting Generative AI

1. GenAI Tools & Platform Access Costs

In the dynamic field of generative AI (GenAI), understanding and managing the costs associated with tools and platform access is critical for users, especially in business and professional settings. The pricing structures for these tools and platforms are not only diverse but also subject to frequent changes, necessitating vigilant monitoring and adaptation.

For GenAI tool licensing, options like ChatGPT Plus and Microsoft 365 Copilot offer monthly subscriptions at $20 and $30 per user, respectively. These subscriptions are tailored to different user bases, with Microsoft's offering being specific to its Microsoft 365 E3, E5, Business Standard, and Business Premium customers.

Commercial AI platforms present a more variable pricing model. OpenAI's GPT-4 Turbo adopts a pay-as-you-go approach, with costs starting at $0.01 per 1,000 tokens for input and $0.03 per 1,000 tokens for output, allowing users to scale their usage according to their needs. Google's AI Vertex Platform, on the other hand, varies its pricing based on usage and model size, starting at $3.75 per hour for training and $0.45 per hour for prediction. This model reflects the resource-intensive nature of AI training and prediction tasks.

The scenario becomes more complex with 3rd party API access. OpenAI's Assistant API, for instance, charges $0.20/GB per assistant per day for retrieval, illustrating a usage-based cost structure. Similarly, Google AI Vertex Marketplace and Microsoft Azure AI Gallery offer variable pricing for their pre-trained models and datasets, catering to a wide range of AI applications. Hugging Face, known for its model hub, bases its pricing on the number of inference requests, underscoring the pay-per-use model prevalent in the industry.

Given the intricate and evolving nature of these pricing structures, continuous and proactive monitoring is essential. Users are advised to stay abreast of the latest pricing updates, regularly review their usage patterns, and adjust their budgets as needed. This approach is critical not only to avoid budget overruns but also to make informed decisions about AI tool selection and usage, ensuring an optimal balance between cost and benefit. Setting up usage thresholds and alerts can significantly aid in this process, offering a safeguard against unexpected expenses and helping in efficient resource management.

2. Prompt Engineering Costs

Prompt engineering is a key component in effectively utilizing Generative AI (GenAI) technologies. It entails the development of precise queries that direct AI models to generate specific desired outputs, making it an indispensable skill in the AI-centric business environment.

- **Prompt Tools and Template Libraries**: This involves investing in sophisticated software and libraries designed to aid in the creation of prompts. The costs for these tools can vary widely, reflecting a range of complexities and functionalities they offer. Businesses might choose from basic packages for simple prompt generation to advanced systems with extensive libraries and customization capabilities.

- **Specialist Support**: The engagement of prompt engineering experts is another cost factor. This could mean hiring dedicated prompt engineers or engaging consultants who specialize in crafting bespoke prompts for unique business requirements. These costs might be structured as salaries for in-house staff or as consultancy fees for external experts.

- **Training and Upskilling**: There's a significant investment in training existing personnel in prompt engineering skills. As businesses look towards integrating prompt engineering as a core competency, the cost of upskilling employees becomes vital. This investment not only empowers staff with cutting-edge skills but also ensures a sustainable integration of GenAI technologies in business processes.

For a detailed exploration of this topic, the book "Prompt Design Patterns: Mastering the Art and Science of Prompt Engineering" offers a comprehensive guide. It delves into various aspects of prompt engineering, making it a valuable resource for those looking to gain expertise in this field.

3. Inference Costs

Inference in Generative AI is a crucial process that involves using trained models to generate outputs based on new inputs. This process is at the heart of GenAI applications, enabling them to provide real-time, responsive services in areas like language processing and image generation.

The inference cost is a significant consideration in the deployment of GenAI. It encompasses the expense associated with calling a Large Language Model (LLM) like GPT-4 or an image generation model such as DALL-E 3.

> **Cost Example:** The cost of using OpenAI's GPT-4 model is based on the number of tokens processed, with the standard GPT-4 model priced at \$0.03 per 1,000 tokens for input and \$0.06 per 1,000 tokens for output. For the larger gpt-4–32k model, the costs are higher, at \$0.06 for input and \$0.12 per 1,000 tokens for output. A token in this context refers to a piece of a word, with 1,000 tokens approximately equal to 750 words, which serves as a measure for the computational resources used.

These costs, driven by the computational power required to process these models, especially the GPU-accelerated servers, can be a significant barrier to adoption for businesses that require the generation of large volumes of content. The infrastructure costs for supporting such high-performance servers, along with the necessary energy costs, are projected to be substantial, potentially exceeding \$76 billion by 2028.

To mitigate these high costs, businesses can explore several strategies. Using smaller models can be a more cost-effective option, though it might come at the cost of reduced capabilities or accuracy. Hosting Open-Source LLMs is another avenue that could offer cost advantages. Additionally, optimizing the inference process itself can lead to more efficient use of resources, potentially lowering costs.

In response to the growing demands of GenAI applications, companies like NVIDIA have developed specialized inference platforms. These platforms, such as the NVIDIA Ada, Hopper, and Grace Hopper processors, are optimized for various generative AI workloads, including AI video, image generation, and large language model deployment. This innovation in inference platforms is critical in enabling the efficient and cost-effective deployment of GenAI applications across various industries.

As GenAI continues to evolve and grow, understanding and managing these costs will be crucial for businesses looking to leverage these technologies effectively. Leaders must consider these factors carefully when planning their GenAI strategies to ensure they can capitalize on the benefits of AI while managing the associated costs.

4. Fine-Tuning Costs

Fine-tuning in Generative AI is essential for businesses as it adapts pre-trained models to specific tasks or domains, enhancing their relevance and effectiveness for particular applications. This process involves updating these models with new data tailored to the unique requirements of the task at hand.

Fine-tuning in Generative AI is a process crucial for adapting pre-trained models to specific tasks or domains. It involves training these models on a new dataset pertinent to the desired output, considering factors such as the model's size and complexity, the amount of data used for fine-tuning, and the number of training epochs.

The cost of fine-tuning GenAI models is influenced by several factors such as the model's size and complexity, the amount of data used for fine-tuning, and the number of training epochs (*A training epoch in machine learning is a single pass through the entire dataset used for training a model*).

While specific real-world cost examples are not readily available, it's important to note that these costs can be significant, especially for complex tasks. Innovations in technology and new platforms, such as Anyscale Endpoints, are emerging to make fine-tuning more cost-effective.

5. Infrastructure Costs

A robust infrastructure is not just a supporting element but a cornerstone in the successful implementation of Generative AI in business. As GenAI technologies advance, the underlying infrastructure must not only match their operational demands but also offer scalability and adaptability to future advancements.

- **Cloud Expenses**: Assessing cloud expenses involves more than just hosting costs. It requires a holistic understanding of the cloud architecture, especially in sectors with sensitive data requirements, such as healthcare. Selecting the appropriate Cloud setup — be it public, private, or multi-cloud — is crucial. This decision must be aligned with the unique demands and scalability needs of GenAI applications.

- **Legacy System Adaptations:** The integration of GenAI with existing legacy systems often necessitates significant modifications, impacting both functionality and financial planning.

- **High Computational Requirements:** The computationally intensive nature of GenAI models calls for substantial investment in enhanced cloud resources or dedicated data centers.

- **Integration and Deployment Costs:** Transitioning GenAI projects from experimental stages to production-ready solutions involves extensive IT infrastructure investments, including advanced compute power, scalable storage solutions, and efficient deployment processes.

Building a strong and flexible infrastructure is essential for businesses to harness the full potential of GenAI. It's a strategic investment that goes beyond immediate technological needs, laying the foundation for future innovation and growth in the era of AI-driven transformation.

6. Data Management Costs

Successfully powering generative AI hinges on robust data management, necessitating significant investment in various areas. When considering the TCO for GenAI, leaders should factor in several critical expenses:

- **Data Storage:** Upgrading storage capabilities is crucial, whether it's expanding on-premises data lakes or enhancing Cloud object stores. These upgrades need to handle large-scale capacities to store source content, labeling workflows, and versioned training sets for models.

- **Data Engineering:** Effective data engineering involves tools and processes for Extract, Transform, Load (ETL) operations, labeling, and machine learning pipelines. These are essential to prepare vast datasets for model training and consumption.

- **External Data Licensing:** Accessing specialized data corpuses, such as scientific publications or creative works, often involves subscription costs and ongoing usage fees for platforms hosting these datasets.

- **Human Review:** Employing subject matter experts to annotate data is critical for ensuring the data aligns with model objectives. This process can be time-consuming and requires a combination of manual effort and assisting tools.

- **Monitoring & Compliance:** For industries like finance and healthcare, maintaining audit trails, conducting quality checks, and performing bias testing are vital to meet compliance and ethical standards.

- **Model Versioning:** Each iteration of model retraining generates new training datasets, necessitating robust version control and lineage tracking systems.

- **Ongoing New Source Ingestion:** Continuously integrating new data sources — from web content to customer engagement data — is key to keeping the model relevant and effective.

Incorporating these expenses into TCO models provides a comprehensive view of the investments required for successful GenAI deployment. It's essential for organizations to recognize these costs early in their AI adoption journey to ensure efficient resource allocation and minimize potential hurdles.

7. Operations Costs

Once live, keeping generative AI solutions humming requires meticulous model operations oversight and workflow automation — spanning data monitoring, retraining cadences, dependency tracking and more. Investing in **MLOps** (Machine Learning Operations) and **AIOps** (Artificial Intelligence for IT Operations) pays dividends optimizing reliability at scale.

- **Continuous Learning and Retraining Costs:** Generative AI models continuously evolve as they encounter new data, necessitating regular retraining. This ongoing process requires a balance between cloud resource utilization and the frequency of updates, which can significantly impact operational costs.

- **Monitoring and Maintenance Expenses:** Effective monitoring of generative AI involves tracking various factors like data drift, model performance, and compliance with ethical standards. Implementing complex dashboards, validators, and manual review processes incurs costs related to personnel and software licenses.

- **Model Drift and Vigilance:** Even stable models are subject to gradual degradation over time due to changes in external factors. Deciding when to retrain models versus adjusting inputs is a critical, ongoing task that requires vigilance and resources.

- **Balancing Speed and Control:** MLOps and AIOps tools provide necessary guardrails and automation, facilitating swift and reliable AI operations. While these platforms optimize operational efficiency, they also require investment in specialized tools, integrations, and organizational change management.

Operational costs associated with MLOps and AIOps are significant but essential for the long-term success and scalability of generative AI projects. Budgeting for these costs is a critical aspect of planning and executing AI strategies, ensuring both operational reliability and efficiency.

8. AI Regulations Compliance Costs

Incorporating the principles of responsible AI into organizational practices not only aligns with ethical standards but also involves significant compliance costs. Understanding and managing these costs is crucial for businesses as they navigate the regulatory landscape of AI.

- **Cost of Transparency Compliance:** Ensuring AI systems are transparent and explainable necessitates investment in technologies and processes that can clarify how AI models make decisions. This might involve developing more interpretable models or acquiring tools that can provide insights into AI processes, which can be resource intensive.

- **Fairness and Bias Mitigation Expenses:** To avoid discrimination and ensure fairness, companies must invest in bias detection and mitigation tools. This includes costs for auditing AI systems for biases, training staff on fairness principles, and potentially redesigning AI systems that exhibit biased behavior.

- **Privacy and Security Investments:** Complying with privacy and security regulations requires robust data protection measures. This can include enhancing cybersecurity defenses, employing encryption techniques, and regularly updating privacy protocols, all of which involve ongoing financial commitments.

- **Accountability and Responsibility Measures:** Establishing clear lines of accountability for AI actions necessitates legal and ethical expertise, as well as mechanisms for monitoring and auditing AI systems. This can lead to additional costs in legal services, insurance, and compliance management.

- **Sustainability-Related Expenditures:** Aligning AI practices with environmental sustainability goals may require investing in energy-efficient AI technologies, which can have higher upfront costs but offer long-term savings and compliance benefits.

- **Human-Centric AI Development Costs:** Ensuring AI augments human capabilities may involve user experience research, human-in-the-loop system designs, and continuous feedback mechanisms to keep AI aligned with human needs and values.

- **Regulatory Compliance Costs:** Adhering to existing and emerging AI regulations requires ongoing monitoring, legal consultation, and adaptation to regulatory changes. This includes expenses for compliance officers, legal advisors, and technology updates to meet regulatory standards.

As AI regulation accelerates, balancing these costs is essential for the successful and responsible deployment of AI technologies. Early investment in compliance can future-proof AI applications, ensuring they meet evolving regulatory standards and ethical considerations in this new era of human-AI collaboration.

9. Talent Costs

In the era of Generative AI, talent acquisition and development are critical components of a successful business strategy. Organizational leaders must navigate challenges of the burgeoning AI talent war and associated costs, while also considering the long-term implications of GenAI on the workforce.

- **Hiring AI Leaders:** The first step in AI-driven business transformation is to hire leaders who can define and implement an effective AI strategy. This involves significant investment but is crucial for guiding the organization's AI journey.

- **Balancing Immediate Talent Needs with Long-Term Strategy:** As GenAI continues to revolutionize various roles and create new job categories, leaders must avoid the pitfalls of a short-term talent rush, which can lead to inflated costs. Strategic planning for medium and long-term talent acquisition is essential to adapt to the evolving landscape without escalating expenses unnecessarily.

- **Leadership and Cultural Development:** The transformative impact of GenAI on business structures and methodologies demands leaders who can guide these changes. Investing in leadership development, both through external hiring and internal training, is crucial. This also involves nurturing a culture that is adaptable to GenAI innovations, which is a significant, yet necessary, cost consideration.

- **The AI Talent War:** With high demand for AI expertise, the competition for top talent is intense. This 'talent war' can drive up salaries and recruitment costs. Organizations must weigh the benefits of attracting external experts against costs and consider alternative strategies like upskilling existing staff.

- **Upskilling and Reskilling as Cost-Effective Options:** Developing internal talent through upskilling and reskilling programs can be a more viable and cost-effective approach to building AI capabilities. This not only helps in retaining employees but also ensures a workforce that is equipped for the AI era.

- **Remote Work Considerations:** Adapting to the preferred working conditions of GenAI professionals, such as remote work, might necessitate changes in corporate culture and infrastructure. Implementing a remote-first or hybrid working model can incur costs but is essential for attracting and retaining AI talent.

- **Preparing for the Next-Gen Workforce:** CEOs and CHROs must anticipate the future of work in the AI era. This includes budgeting for the development of new roles and competencies that GenAI will necessitate, ensuring the workforce remains relevant and competitive.

The talent costs associated with GenAI are multifaceted and require careful strategic planning. Balancing the immediate need for AI talent with long-term workforce development, while managing the costs associated with hiring, upskilling, and cultural transformation, is key to thriving in the AI-driven business landscape.

Cost Type	Description	Example Components
1. GenAI Tools & Platform Access Costs	Expenses related to accessing generative AI tools and commercial platforms	• GenAI tool licensing fees e.g. ChatGPT Plus at $20/month per user • Commercial AI platform subscription fees e.g. Anthropic, Cohere, etc. • 3rd party API usage fees
2. Prompt Engineering Costs	Investments needed for developing prompts to guide AI systems	• Prompt engineering tools and libraries • Hiring specialists or consultants • Training employees on prompt engineering
3. Inference Costs	Fees connected to processing inputs and generating outputs from AI models	• Per token pricing models e.g. $0.03 to $0.12 per 1000 tokens for GPT-4 • Infrastructure for high performance computing servers • Specialized AI inference platforms like NVIDIA Hopper
4. Fine-Tuning Costs	Expenses related to customizing or adapting generic AI models	• Model training on domain-specific datasets • Number of training epochs • Emerging fine-tuning platforms such as Anyscale Endpoints
5. Infrastructure Costs	Costs stemming from IT infrastructure to support AI systems	• Cloud expenses • Legacy systems integration • Computational requirements • Deployment and management tools
6. Data Management Costs	Investments needed for ingesting, preparing, labeling, storing and monitoring AI training data	• Data storage capacity • Data engineering pipelines • Manual data review and labeling • Monitoring tools and model versioning
7. Operations Costs	Regular expenses needed to maintain and operate AI systems	• Continuous model retraining cycles • Monitoring, validation and maintenance tools • Addressing model drift • MLOps software and integrations
8. AI Regulations Compliance Costs	Costs arising from complying with laws around ethical AI system practices	• Transparency and explainability tools • Fairness testing and bias mitigation • Privacy and security measures • Legal, insurance and compliance management
9. Talent Costs	Investments connected to building an AI-ready workforce	• Salaries for AI leadership and specialists • Upskilling programs • Remote work policy changes • Preparing workforce for evolving roles

Table 5-1: The AI TCO Cost Types, Descriptions, and Example Components

Final Reflections

As we delve into the rapidly shifting world of generative AI, business leaders find themselves at the helm of a complex voyage. They are faced with the task of weighing the promising benefits against the inherent risks and costs associated with these advanced technologies. This guide has endeavored to provide a comprehensive examination of the various elements that form the Total Cost of Ownership (TCO) in deploying generative AI. It spans a broad spectrum, addressing aspects from accessing foundational models to maintaining operational compliance, and managing both operations and talent effectively.

Grasping and quantifying the multifaceted costs involved in the adoption of generative AI is an intricate yet vital undertaking. It demands a detailed consideration of the long-term investments essential for a responsible and effective implementation. This includes preparing for the expenses related to core model usage, recruiting specialized talent, establishing a strong data infrastructure, and committing to ongoing model retraining. Additionally, the financial implications of ensuring regulatory compliance and managing the impacts of organizational change are equally significant.

The insights offered in this guide aim to provide executives and innovators with a strategic framework. This framework is designed to assist in careful budget assessment, informed decision-making regarding platform choices and staffing, and the effective planning of deployment phases. Although the cost trajectory of generative AI is dynamic and sometimes unpredictable, leaders who approach these challenges with strategic foresight can optimize returns and foster lasting business transformation.

With meticulous planning, strategic implementation, and a foundation in solid principles, the escalating TCO of generative AI can be channeled towards significant, enduring returns on investment (ROI). While the initial costs and complexities of adoption may seem daunting, the potential for industry reinvention and growth is concurrently expanding. For those who strategically embark on this journey, the value yielded can far surpass the initial investments, paving the way for substantial, transformative outcomes in their respective industries.

Chapter 6

AI ROI and ROE Analysis

"We tend to overvalue the things we can measure and undervalue the things we cannot." — John Hayes

I n the dynamic realm of business technology, where Generative AI (GenAI) is rapidly redefining the status quo, comprehending its multifaceted impact is essential for modern organizations. This chapter serves as a crucial guide, delving into the complexities of integrating GenAI, with a focus on understanding its Return on Investment (ROI), Return on Experience (ROE), and Total Cost of Ownership (TCO). It builds upon the insights provided in the previous two chapters, which meticulously explored the business value of GenAI and a detailed analysis of TCO, setting the stage for a thorough examination of ROI.

Beginning with an in-depth look at ROI, the chapter navigates through the financial intricacies of GenAI, highlighting how it influences operational efficiencies and financial outcomes. Moving beyond mere numbers, the chapter then shifts its focus to the human element of technology, exploring the ROE. This section sheds light on how GenAI reshapes organizational culture, enhances employee engagement, and elevates customer experiences.

In parallel, the narrative weaves in the critical aspect of TCO, drawing from the groundwork laid in the preceding chapter. This comprehensive view encapsulates the entire investment landscape of GenAI, from initial deployment costs to long-term financial commitments, offering a well-rounded perspective on the total financial implications of GenAI initiatives.

This chapter, by juxtaposing ROI, ROE, and TCO, provides a balanced and holistic understanding of GenAI's impact. It equips business leaders, strategists, and decision-makers with the necessary insights to navigate the GenAI revolution effectively. The aim is to ensure that GenAI integration is not just technologically advanced but also strategically aligned with the broader objectives and values of the organization. This approach paves the way for businesses to harness the full potential of GenAI, making informed decisions that foster both operational excellence and sustainable growth.

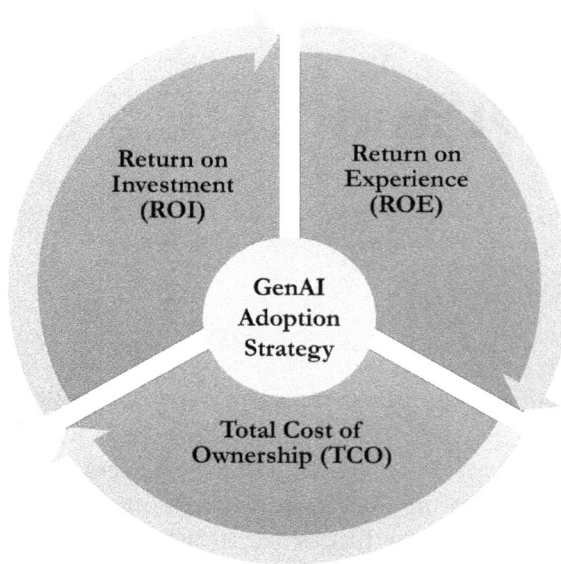

Figure 6-1: ROI, ROE, and TCO in GenAI Adoption Strategy

AI Return on Investment (ROI) Analysis

In the transformative era of Generative AI (GenAI), understanding and calculating Return on Investment (ROI) has become more complex and multifaceted. Business leaders and strategists face the dual challenge of not only tapping into the vast potential of GenAI but also quantifying its value in concrete, tangible terms. Determining the ROI of GenAI initiatives goes beyond simple calculations, influenced by a variety of factors including the breadth of implementation, the specific business goals driving the technology's integration, and the timeline over which it is deployed.

In this complex landscape, it is crucial for leaders to undertake a comprehensive analysis of the different aspects of GenAI integration. Developing realistic and insightful ROI projections requires a deep understanding of GenAI's impact not just in terms of immediate financial returns, but also its broader influence on the organization's operational efficiency, competitive standing, and potential for long-term growth. This approach necessitates a look at the key considerations for GenAI ROI, which involve more than just the surface-level financial metrics, delving into the strategic implications of GenAI adoption across the business.

1. ROI Dynamics of GenAI: Understanding Scope and Scale

The integration of Generative AI (GenAI) in business operations brings forward the critical concepts of 'Implementation Scope' and 'Scale', which are instrumental in determining the Return on Investment (ROI).

Implementation Scope in GenAI refers to the breadth and depth of its integration within an organization. This can range from deploying specific applications like automated customer service tools to implementing GenAI in a comprehensive, cross-departmental manner. The scope essentially dictates how widely and deeply GenAI is woven into the organizational fabric.

Implementation Scale deals with the extent or magnitude of GenAI deployment. This varies from limited departmental applications to full-scale, enterprise-wide integration, reflecting the overall reach of GenAI within the company.

Understanding the intersection of scope and scale is crucial as it directly influences the complexity, resource allocation, and the timeline for realizing ROI from GenAI projects.

Diverse Implementation Types and Their Impacts

GenAI implementations can be categorized into three types, each with typical timeframes for ROI realization:

1. **Narrow Implementations:**

    ○ **Focus:** Specific tasks such as customer service chatbots or automated content creation tools.

    ○ **Advantages:** Lower cost, quicker ROI, ease of integration.

    ○ **Typical Timeframe for ROI:** 6-12 months, ideal for targeted improvements and cost reductions.

2. **Broad Implementations:**

    ○ **Focus:** Application across multiple departments or processes, like in supply chain management and HR.

    ○ **Advantages:** Potential for transformative impacts and process optimization.

○ **Typical Timeframe for ROI:** 1-2 years, reflecting the substantial investment and larger scope of integration.

3. **Organizational Transformations:**

○ **Focus:** Embedding GenAI into the core of business operations and strategies.

○ **Advantages:** Long-term competitive advantages and potential for significant business model shifts.

○ **Typical Timeframe for ROI:** 2-3+ years, due to substantial upfront investment, strategic planning, and the transformative nature of these projects.

Key Factors Influencing Scope and Scale

In the landscape of Generative AI (GenAI) implementation, several interconnected factors play a pivotal role in shaping the decision on its scope and scale. The specific challenges and strategic goals of an organization are at the forefront of these considerations. They dictate the extent to which GenAI can be adopted, ensuring that the technology aligns with and supports the organization's overarching ambitions and objectives.

The availability of resources, encompassing financial, human, and technical assets, sets the practical boundaries for GenAI implementation. An organization's capacity to invest in GenAI, the expertise of its workforce, and the technological infrastructure available are key determinants of how far and how deep GenAI can be integrated into its processes.

Technological maturity also holds significant sway in this decision-making process. The readiness and sophistication of available GenAI solutions can either enable or limit their deployment. As technology evolves, it opens new avenues for application, allowing organizations to push the boundaries of what's possible with GenAI.

Market dynamics constitute another critical factor. The external environment, including market trends and competitive pressures, can influence organizations to adopt more aggressive or conservative GenAI integration strategies. Staying responsive to these external factors ensures that an organization remains competitive and relevant in its industry.

Strategic Considerations for Effective Implementation

Effectively implementing GenAI is akin to performing a careful balancing act. One of the primary considerations is risk management, especially for larger GenAI projects. These initiatives come with inherent risks, requiring meticulous planning and robust change management strategies to mitigate potential pitfalls.

Adaptability is another key aspect. As GenAI usage within an organization grows and evolves, the ability to adapt to new technologies and processes becomes crucial. This flexibility ensures that the organization can maximize the benefits of GenAI while navigating the challenges that come with technological integration.

Perhaps the most crucial element in this balancing act is aligning immediate needs with the long-term vision. It's essential for organizations to marry their immediate ROI expectations with their long-term strategic objectives. This alignment ensures that GenAI initiatives are not just successful in the short term but also contribute to the organization's enduring growth and evolution.

In summary, the decision regarding the scope and scale of GenAI implementation is a complex yet crucial one, as it shapes the potential ROI. It requires a nuanced understanding of an organization's objectives, capabilities, and external environment. Managing these aspects effectively is key to maximizing the benefits of GenAI, ensuring it aligns with the company's overarching strategy and capacity for innovation and change.

2. Impact of Custom vs. Prebuilt GenAI Models on ROI

In the realm of Generative AI (GenAI) integration, the choice between custom and prebuilt models is a pivotal one, significantly influencing both the initial investment and the long-term strategic impact on Return on Investment (ROI).

Prebuilt GenAI Models are often seen as the more cost-efficient route. They typically require lower upfront investments, sometimes up to 30% less than their custom counterparts. This aspect makes them attractive for organizations seeking quick deployment and faster ROI. However, the trade-off comes in the form of limited strategic impact. Given their widespread availability, prebuilt models offer limited competitive differentiation, which might not be fully aligned with specific business needs or strategic goals.

On the other hand, **Custom GenAI Models** demand greater initial development efforts, leading to higher upfront costs. The long-term benefits of these models are pronounced in their ability to yield more significant strategic impacts and competitive advantage. These models are tailored to align with unique business goals and processes, offering a more profound integration with the organization's operational fabric. The ROI for custom models generally materializes over a longer period, typically between 24 to 36 months, due to the initial development phase and the time required for effective integration into business processes.

The decision between adopting custom or prebuilt models hinges on an organization's strategic priorities. If the immediate goal is quick wins and efficiency gains, prebuilt models are more apt. Conversely, for organizations aiming for long-term value creation and a sustained competitive edge, investing in custom models becomes more advantageous. This choice also depends on the organization's capacity to invest in and support the development of custom solutions, both financially and technically.

The selection between custom and prebuilt GenAI models directly impacts the ROI, affecting both the speed of realization and the depth of strategic value. Prebuilt models offer a quicker and more cost-effective solution for immediate needs, while custom models provide greater long-term benefits and strategic alignment. This critical decision reflects an organization's immediate operational needs and its long-term strategic ambitions.

3. Milestones and Metrics: Refining ROI Calculations in GenAI

In the realm of Generative AI (GenAI), the approach to calculating Return on Investment (ROI) needs to be multifaceted, encompassing a range of metrics that go beyond traditional efficiency and cost measures. It's essential to incorporate engagement indicators such as customer satisfaction, employee experience, and stakeholder confidence. These broader metrics offer a more comprehensive view of GenAI's impact.

The process of setting precise milestones is critical for tracking and projecting ROI in GenAI initiatives. These milestones should be closely aligned with the specific goals of each use case. For example, in a manufacturing context, key metrics might include production output and downtime. These measures provide direct insight into operational efficiency and the impact of GenAI on production processes. In contrast, for HR applications, the focus would be on metrics like recruitment efficiency and staff retention rates, offering a view of how GenAI influences human resources management.

The projection of AI ROI demands a holistic approach. This entails setting clear milestones that tie together both efficiency and cost metrics across the various business operations and processes impacted by the GenAI implementation. These metrics should also be supported by relevant Key Performance Indicators (KPIs). The goal here is not merely to assess cost savings but to also gauge productivity gains, quality improvements, customer satisfaction levels, revenue increases, and risk mitigation.

Establishing measurable milestones that are in sync with the goals of specific use cases enables more accurate tracking and assessment of AI ROI. This approach ensures that the ROI calculation captures the full spectrum of GenAI's impact on the organization, from operational enhancements to broader strategic gains. By doing so, businesses can gain a clearer understanding of the value GenAI brings, guiding more informed decision-making and strategy formulation.

4. Assessing GenAI's Cross-Functional ROI Impact

When evaluating the Return on Investment (ROI) of Generative AI (GenAI) initiatives, it is crucial to consider their cross-functional impact. GenAI's influence often extends beyond its immediate area of application, affecting a range of interconnected business processes. This broader impact necessitates an ROI assessment that spans multiple domains within the organization.

Take, for instance, the implementation of GenAI in supply chain management. Such an integration does not solely affect the supply chain operations; it also has a significant impact on related areas like manufacturing schedules and inventory control. The introduction of GenAI in forecasting demand can lead to more efficient manufacturing schedules, reducing downtime and optimizing production cycles. Similarly, improved inventory control through GenAI can result in cost savings and reduced waste, contributing positively to the overall ROI.

This example underscores the need for an ROI assessment approach that captures the full scope of GenAI's impact. Assessing ROI across multiple domains allows for a more comprehensive understanding of GenAI's value to the organization. It enables businesses to appreciate how changes in one area, influenced by GenAI, can lead to benefits or challenges in another, ensuring a holistic view of the investment's effectiveness.

In essence, a cross-functional perspective in ROI assessments is essential to accurately gauge the wide-ranging effects of GenAI. This approach ensures that businesses can make informed decisions about GenAI investments, fully understanding their potential to drive efficiency and innovation across various facets of the organization.

5. Value Realization: Turning GenAI Capabilities into Tangible Returns

In the journey of leveraging Generative AI (GenAI) for business enhancement, the steps taken after deployment play a pivotal role in translating GenAI capabilities into tangible financial returns. It's not just about harnessing the advanced capabilities of GenAI but also about effectively applying these insights to drive real-world business outcomes.

A key aspect of this post-deployment phase is the conversion of generative customer insights into actionable marketing strategies. GenAI can offer deep insights into customer behaviors, preferences, and trends. However, the true measure of ROI is realized when these insights are effectively transformed into marketing strategies that resonate with the target audience, leading to more engaging and successful campaigns.

The overall ROI from GenAI, therefore, hinges not just on the technology's deployment but significantly on how well these advanced insights are executed in practical business scenarios. It involves a strategic shift from merely collecting and analyzing data to actively applying this knowledge in creating more impactful and result-oriented marketing campaigns.

This step in the GenAI integration process underlines the importance of not just technological proficiency but also strategic acumen in realizing the full value of GenAI investments. It's about bridging the gap between advanced AI capabilities and their practical application in enhancing business processes, particularly in customer engagement and marketing effectiveness.

In essence, the value realization phase is crucial in determining the success and ROI of GenAI initiatives. It's where the advanced capabilities of GenAI are translated into concrete, measurable business outcomes, reaffirming the technology's role as a driver of modern business transformation.

The Future of ROI in the Age of Generative AI

As the landscape of business and technology continues to evolve with the advent of Generative AI (GenAI), the approach to calculating and understanding Return on Investment (ROI) also undergoes a significant transformation. In this new era, the ROI of GenAI is not just a numerical value but a comprehensive measure of how this groundbreaking technology reshapes and enhances various facets of business operations.

Business leaders and strategists must look beyond the conventional metrics and embrace a broader perspective that includes the multifaceted impacts of GenAI. From improving operational efficiency to redefining competitive dynamics and fostering long-term growth, the implications of GenAI integration are vast and far-reaching. The challenge lies in effectively harnessing this potential and translating it into quantifiable outcomes.

The journey to a successful GenAI implementation and its ROI realization involves meticulous planning, strategic integration, and the alignment of technology with business objectives. This process requires an in-depth understanding of the technology's capabilities, a clear vision of the desired outcomes, and a commitment to adapting and evolving with the changing technological landscape.

In conclusion, the ROI analysis of GenAI is a dynamic and ongoing process, demanding continuous learning and adaptation. As organizations navigate this journey, the key to success lies in their ability to balance immediate gains with long-term strategic goals, ensuring that GenAI becomes a powerful tool in their arsenal for achieving sustainable business growth and innovation. The era of GenAI presents not just challenges but also unprecedented opportunities to redefine the value and impact of investments in technology.

Beyond ROI: Embracing ROE in GenAI

In the evolving landscape of Generative AI (GenAI), the traditional focus on Return on Investment (ROI) is increasingly complemented by the equally important concept of Return on Experience (ROE). While ROI provides a quantitative measure of financial success, ROE brings in a qualitative dimension, capturing the broader impacts on experiences across the organization — from employees to customers and beyond. This dual focus ensures a more holistic assessment of GenAI's value.

- **ROE in Employee Engagement and Satisfaction:** GenAI's role in automating routine tasks not only boosts operational efficiency, contributing to ROI, but also significantly enhances employee experiences. This elevation in job satisfaction and engagement, a key aspect of ROE, leads to a more motivated and innovative workforce, indirectly contributing to the overall ROI through increased productivity and creativity.

- **ROE in Customer Experience:** The impact of GenAI on customer interactions and satisfaction is a crucial part of ROE. By improving the quality of customer engagements and personalizing experiences, GenAI enhances customer loyalty and retention, which, in turn, positively influences ROI through sustained revenue streams and potential market growth.

- **Operational Efficiency and User Interaction:** While operational efficiency directly feeds into ROI calculations, the improvements in user interactions and interfaces also contribute significantly to ROE. Better user experiences lead to increased adoption and positive perceptions of the organization, which indirectly support long-term ROI.

- **Innovation and Creativity:** The liberation of employees from repetitive tasks opens doors to innovation, a key component of ROE. This fosters an environment conducive to generating new ideas and solutions, which can lead to the development of new revenue streams, again complementing ROI.

- **Data-Driven Decision Making:** ROE encompasses the empowerment brought about by GenAI in making informed decisions. This not only enhances the strategic capabilities of an organization but also supports ROI by enabling more effective and efficient decision-making processes.

- **Work-Life Balance:** GenAI's role in facilitating flexible work models contributes significantly to ROE by improving work-life balance and employee well-being. This aspect, while not directly quantifiable in ROI terms, contributes to long-term organizational success by reducing turnover and attracting top talent.

The integration of ROE into the assessment of GenAI initiatives provides a more comprehensive understanding of the technology's impact. While ROI remains a crucial metric, ROE offers insights into the qualitative enhancements that GenAI brings to an organization. Together, ROI and ROE present a complete picture of GenAI's value, encompassing both its financial benefits and its transformative impact on various aspects of the business environment. In the age of GenAI, recognizing and emphasizing the importance of ROE alongside ROI is essential for a full appreciation of the technology's potential to drive sustainable growth and success.

Industry-Specific Considerations in GenAI's ROI and ROE

The timelines for achieving Return on Investment (ROI) and the nature of Return on Experience (ROE) benefits from Generative AI (GenAI) demonstrate significant variability across different industries. This variation is deeply rooted in the distinct characteristics and specific needs unique to each sector, which fundamentally shape the adoption of GenAI and the resulting benefits.

- **Varied ROI Timeframes:** In industries like retail or customer service, GenAI can quickly streamline processes and enhance customer interactions, leading to a rapid ROI. However, in sectors such as healthcare or aerospace, where GenAI applications might involve more complex, safety-critical systems, the ROI timeframe can be significantly longer due to stringent testing and regulatory approval processes.

- **ROE in High-Tech vs. Traditional Sectors:** The impact of GenAI on employee experience (ROE) also differs by industry. In high-tech sectors, GenAI can rapidly enhance creativity and innovation, leading to high ROE through improved job satisfaction and efficiency. In more traditional industries, the introduction of GenAI might require substantial cultural shifts and training, with ROE realized over a longer period.

- **Customization and Specialization Needs:** Different industries have varying degrees of need for customization in GenAI applications. Highly specialized sectors may require significant investment in developing bespoke AI solutions, impacting both ROI and ROE.

- **Compliance and Ethical Implications:** Industries such as finance and healthcare have strict regulatory and ethical standards, which can influence the costs and complexity of GenAI deployment. Ensuring compliance and ethical integrity in these sectors may require additional resources, affecting ROI calculations.

- **Operational Efficiency vs. Innovation Focus:** Some industries might prioritize GenAI for operational efficiency, leading to direct cost savings and a clear ROI. Others may focus on leveraging GenAI for innovation and market differentiation, where the ROI is more qualitative and linked to long-term market positioning.

Recognizing these industry-specific factors is crucial in determining the ROI and ROE of GenAI. A nuanced understanding of these elements is vital for businesses aiming to develop GenAI strategies that are not just technologically sound but also perfectly aligned with their industry's unique characteristics, regulatory requirements, and long-term goals.

This tailored approach ensures that GenAI deployment is not a one-size-fits-all solution but a strategic decision shaped by the specific contours of each industry. By doing so, businesses can maximize the potential of GenAI, ensuring that it contributes effectively to both immediate operational efficiency and long-term strategic objectives.

Integrating Key Metrics in GenAI Strategy: ROI, ROE, and TCO

For executives planning to integrate Generative AI (GenAI), it's essential to adopt a comprehensive approach that includes Return on Investment (ROI), Return on Experience (ROE), and Total Cost of Ownership (TCO). This holistic strategy provides a full picture of the impact and practicality of GenAI technologies.

Focusing on ROI involves quantifying the direct financial benefits GenAI brings. This includes assessing cost savings, revenue enhancements, and overall profitability from GenAI integrations. It's crucial to outline the financial trade-offs and identify the payback periods for various GenAI adoption scenarios, offering a clear understanding of the financial returns.

Alongside ROI, evaluating ROE is pivotal. ROE examines GenAI's impact on organizational culture, particularly how it influences employee experiences by boosting engagement, productivity, and satisfaction. This metric assesses the more qualitative aspects of GenAI, underscoring its potential to transform workplace dynamics and foster creativity.

A thorough TCO analysis complements these perspectives. It encompasses all costs related to developing, deploying, and maintaining GenAI solutions. This includes not just direct expenses but also those often overlooked, such as investment in talent, infrastructure, system integration, and regulatory compliance. TCO gives a comprehensive view of the long-term financial commitment involved in GenAI initiatives.

By simultaneously considering ROI, ROE, and TCO, executives can gain insights into the tangible returns and cultural shifts introduced by GenAI. These multifaceted assessments provide the financial and human-centric foresight needed to formulate sophisticated and balanced GenAI adoption strategies. Armed with this comprehensive understanding, organizations can leverage GenAI effectively, enhancing operational efficiency and human potential.

Chapter 7

AI Risk Mitigation
Framework

"If you don't invest in risk management, it doesn't matter what business you're in, it's a risky business." — *Gary Cohn*

I n the realm of artificial intelligence, the glittering promise of innovation often overshadows a less glamorous reality: a labyrinth of risks and challenges that can undermine the very foundations of trust and reliability upon which these technologies stand. From security vulnerabilities that open the door to data theft, to algorithmic biases that perpetuate social injustices, the pitfalls of AI deployment are as real as they are varied.

But what if these risks could be transformed from looming threats into opportunities for responsible innovation?

This is the guiding question at the heart of our approach to AI risk management. Far from being a mere afterthought or a compliance checklist, our methodology is a proactive journey through the intricacies of AI governance. We delve into the realms of generative AI quality, security, privacy, fairness, transparency, misuse, and regulatory compliance, not just as isolated domains, but as interconnected facets of a comprehensive risk landscape.

From the initial spark of prototyping to the rigorous demands of production monitoring, we rigorously evaluate and mitigate these multifarious risks. Our strategy is not just to react to vulnerabilities, oversights, and ethical dilemmas, but to anticipate and address them head-on, thereby preserving stakeholder trust and sustaining the potential for ongoing innovation.

Our multifaceted risk framework is a dynamic tapestry woven from attack simulations, audits, consensus standards, policies, platforms, and transparent assessments. This framework is not static; it evolves continuously, fortified by insights shared across the AI ecosystem. In this complex dance of productivity and principles, compliance is not just a legal requirement but a stepping stone towards winning the trust of customers and the public.

In the following sections, we will detail how our approach not only upholds safety for all involved but also empowers our core mission through responsible, vigilant innovation. Join us in exploring how proactive risk management is not just a safeguard, but a catalyst for transformative AI advancement.

Risk Category	Risks	Mitigations
Output Quality	Factual inaccuracies, Bias, Hallucinations	Accuracy benchmarking, Input validation, Human-in-loop reviews
Data Security	Breaches and theft, Weak protocols, Unencrypted data, Insufficient auditing	Access controls, Encryption, Activity monitoring
Privacy	Attribute linkage, Dataset bias, Information leakage	Federated learning, Differential privacy, Scrubbing tools
Bias and Fairness	Representation imbalance, Outcome disparities, Historical stereotyping	Bias testing, Synthetic oversampling, Counterstereotypical data
Transparency	Black box opacity, Observability gaps, Data lineage confusion	Local model explanations, Metadata persistence, Clean room techniques
Misuse and Harms	Disinformation, Adversarial attacks, Fraud	Credibility scoring, Anomaly detection, Context watermarking
Compliance	Data sovereignty confusion, Incomplete documentation, Policy violations	Multi-cloud data separation, Model cards, Automated conformance checks

Table 7-1: The AI risks categories, descriptions, and mitigation solutions

1. Output Quality Risks

The reliability, accuracy, and appropriateness of generative AI outputs represent a complex challenge that requires a nuanced approach to mitigate risks across various quality dimensions. A primary concern in this domain is addressing the risk of factual inaccuracy. Generative models, despite their logical coherence, often encounter limitations in producing factually correct outputs due to constrained world knowledge or inherent language modeling biases. This issue is significant: increased exposure to inaccuracies can erode the ability of users to discern truth.

To tackle this, strategies such as continuous accuracy benchmarking against test datasets are employed. This process aids in identifying and flagging deviations, prompting timely model retraining. Enhancing AI reliability also involves implementing input validation measures, such as blacklisting or quarantining unverified claims, thereby requiring additional references before generation. Complementing these technical measures is the human element—a review process involving subject matter experts who act as a final safeguard against inaccuracies.

Another crucial aspect is addressing bias risks. Historical data biases, known for perpetuating stereotypes, often lead to the overrepresentation of majority groups and marginalization of minorities. Counteracting this involves developing bias testing suites that quantify disparities, providing a data-driven basis for model updates aimed at balanced representation. Adversarial debiasing techniques are crucial in removing encoded biases while maintaining task efficacy. Additionally, incorporating diversity-aligned datasets that provide counterexamples to stereotypes enriches the model's understanding and fairness.

Hallucination risks, where AI generates completely fictitious content or imagery, pose another challenge, especially when such outputs could be misleading if perceived as factual. To address this, fake content detectors using forensic media analysis are integrated. These tools initiate manual review processes before any final output is approved. Digital watermarking is also employed to mark synthetic media, clearly indicating its artificial origin. An important strategy involves labeling fictional outputs as imaginary, setting correct user expectations and distinguishing them from unqualified factual claims.

In combating hallucination risks, there's an integration of cutting-edge strategies from recent research. This enhanced approach includes Retrieval-Augmented Generation (RAG), which utilizes external knowledge bases for more accurate and informed responses. A suite of innovative techniques is adopted, spanning the entire generation process: pre-generation tools like LLM-Augmenter, during-generation strategies like Knowledge Retrieval and EVER (Real-time Verification and Rectification), and post-generation methods such as RARR (Retrofit Attribution using Research and Revision). These advancements, along with novel decoding strategies and knowledge graph integration, are revolutionizing AI performance. By incorporating these sophisticated methods, the risk management framework becomes more robust, ensuring that AI systems maintain high standards of accuracy and fairness, vital for executive decision-making.

In conclusion, by continuously assessing and refining the approach to output quality in the domains of accuracy, bias, and fake detection, and supplementing these with robust human oversight procedures, it is ensured that generative models are not only reliable but also adhere to the highest ethical standards. This governance framework empowers generativity while safeguarding against critical flaws.

2. Data Security Risks

Ensuring the security of datasets in generative model development is a critical task, pivotal to averting the theft of sensitive information or proprietary intellectual property. This task becomes increasingly complex with the presence of vulnerabilities that can lead to financial, competitive, and reputational risks. Key data security risks and their mitigations are outlined as follows:

Breaches and theft present a significant threat, with the potential for external attackers or malicious insiders to gain unauthorized access to confidential information. To mitigate this, stringent access controls are essential, limiting internal visibility solely to personnel who require access. Additionally, the implementation of multi-factor authentication provides a layered defense against compromised credentials. Moreover, regular cybersecurity penetration testing is crucial to fortify infrastructure and identify vulnerabilities proactively.

Weak access protocols, characterized by inadequate data access approval workflows or excessive unlabeled access, increase the risk of insider threats and compliance violations. To address this, formal request ticketing procedures are established, documenting the justification for access and requiring oversight approval, thereby preventing unlabeled access. Furthermore, instituting access expiry with mandatory renewal on a quarterly basis helps maintain visibility and control.

The risk of unencrypted data flow, which can lead to exposure opportunities for man-in-the-middle attacks, is another concern. Encrypting data both at rest and in transit is a fundamental mitigation strategy, safeguarding against such exposure. In addition, key management administration is crucial, ensuring secure handling of encryption keys and protecting encrypted data through crypto-shredding.

Insufficient auditing practices, which result in limited visibility into access logs, policy violations, or suspicious queries, hamper effective breach detection. Implementing robust logging with centralized analysis is vital for spotlighting anomalous behavior to information security teams. Integrating daily audit reviews that include user behavior analytics helps in identifying outliers, thereby enhancing security monitoring.

By hardening data security across the data lifecycle and supplementing it with comprehensive oversight, access governance, encryption, and auditing, a robust framework is established. This framework not only ensures responsible data stewardship but also significantly reduces risks associated with AI systems and stakeholders, maintaining a secure and trustworthy environment for generative model development.

3. Privacy Risks

In the realm of data processing, safeguarding individuals' personal information and their right to consent, even in the context of anonymized data, is of paramount importance. This is due to the persistent risks of attribution linkage and pattern inference reverse engineering, which carry legal, ethical, and brand reputation implications. The following elaboration delves into the key privacy risks and their corresponding mitigation strategies:

The risk of attribute linkage arises when anonymized behavioral data is connected with other datasets using common identifiers, potentially revealing identities without explicit consent. To counteract this risk, the use of federated learning is recommended. This approach keeps data partitioned, avoiding pooling into centralized stores. Additionally, employing encryption schemes like homomorphic encryption permits computations on ciphertexts, safeguarding privacy. Furthermore, trusted execution environments offer hardware-based confidential computing, adding another layer of security.

Dataset bias presents another significant risk. This occurs when datasets polluted with discrimination or skewed subgroup representation lead to biased outcomes, failing privacy fairness tests. To mitigate this, proactive bias testing is employed to quantify and address imbalances through strategies like model tuning, reweighting, and synthetic oversampling. Adversarial debiasing techniques are also utilized to alter latent space representations, thereby 'forgetting' prohibited attributes. Expanding diversity in hiring brings new perspectives that ensure balance and fairness in dataset handling.

Information leakage, stemming from residual dataset artifacts incidentally retained due to incomplete sanitation, can enable the reconstruction of identities. To mitigate this, automated scrubbing tools are deployed to purge all predefined personal identifiers, adhering to strict privacy policies. Data protection impact assessments are conducted to highlight residual risks. The application of differential privacy introduces mathematical noise, effectively obscuring individual identities within datasets.

Continuously conducting privacy risk assessments and making necessary adjustments to mitigation strategies is essential. This ensures that AI systems remain productive and responsible, respecting user consent prerogatives in a landscape of evolving technologies and regulations. It's a commitment to maintaining privacy integrity in data processing, crucial for upholding trust and compliance in the digital age.

4. Bias and Fairness Risks

Historical biases present in datasets pose a substantial risk in generative models, leading to discriminatory, prejudicial, or negatively skewed outputs that fail to meet ethical standards across various sensitive attributes such as gender, race, and socioeconomic status. Addressing these core bias and fairness risks involves several key strategies:

Representation imbalance occurs when there is a gender, racial, age, or other group imbalances in the training data, leading to distorted perceptions and the marginalization of minority groups. To mitigate this risk, bias testing suites are utilized to continuously quantify dataset and model disparities on protected attributes. Synthetic oversampling is applied to increase the representation of minority groups in the datasets. Moreover, expanding diversity in hiring brings in new perspectives, balancing teams and data.

Outcome disparities, such as different error rates for different ethnicities or genders, reflect biased assessments in model performance. Parity analysis is an effective tool for identifying unbalanced model performance, prompting the need for tuning on specific subgroups. Adversarial debiasing techniques are employed to remove encoding bias while retaining the efficacy of the task. Additionally, a human-in-the-loop review process is established to check model decisions on protected classes, ensuring fairness.

Historical stereotyping, where generative content reflects real-world stereotypes baked into the training data, poses the risk of reinforcing these biases. Style transfer reconstruction is used to remove stylistic giveaways while retaining meaning, thus avoiding stereotypical rendering. Ingesting counter-stereotypical data helps balance representations in the model. Furthermore, warnings are put in place to alert consumers to potential outdated biases, flagging the need for human review.

By proactively addressing fairness across identity characteristics, these strategies contribute to sustaining equitable AI systems. This commitment is aligned with ethical priorities, recognizing the growing societal responsibilities of algorithmic systems. It's a comprehensive approach to ensuring that AI technologies are developed and deployed in a manner that respects and promotes fairness and equality.

5. Transparency and Explainability Risks

The challenge of deciphering the decision-making processes of generative models – understanding their reasoning and scoring methodologies – poses significant risks to model interpretability, improvement, and accountability. This section discusses the core transparency risks and their mitigation strategies:

The issue of black box opacity arises from the complex inner workings of neural networks, which often defy straightforward explanations. This complexity limits the potential for feedback and refinement. To mitigate this, local interpretable model approximations, such as LIME, are employed to determine the drivers of individual predictions. Representation learning techniques are also used to extract relationships within learned representation spaces. Additionally, the development, testing, and performance of models are documented in model cards, providing crucial information to external stakeholders.

Observability gaps, which refer to the inability to monitor system health metrics or decision factors in real-time, can significantly impede diagnostics. Addressing these gaps involves consistent logging at modular levels to track lineage, data drift, and the effects of retraining on subcomponents. Continuous integration testing automates checks within the pipeline, ensuring consistent performance. Gradual rollout strategies that incorporate observational probes facilitate incremental improvement and troubleshooting.

Data lineage confusion can occur when there's disjointed tracking of dataset sources, contents, and processing, lacking proper annotations. This can risk the reproducibility of models. To counter this, metadata persistence is crucial, cataloging details such as the datasets used, their volumes, features, origin, processing logic, and the intended purpose of the model. Graph databases are utilized to visualize complex pipeline connections, aiding in troubleshooting. Additionally, clean room techniques log steps for reproducibility, including controls for random seed settings.

Enhancing the understanding of how models function and the reasoning behind their actions is crucial for continuous improvement of AI systems. This ensures they operate reliably and avoids potential harms. Maintaining transparent and explainable operations is key to sustaining healthy model hygiene, fostering trust, and ensuring accountability in the evolving landscape of AI technology.

6. Misuse and Harms Risks

Generative models, while innovative, face the risk of being misused by malicious actors. This misuse can manifest in the creation of synthetic disinformation, phishing content for fraud, or adversarial inputs that trigger unsafe system behavior. Addressing these misapplications requires a multi-faceted approach:

Disinformation risks emerge when generative models are used to create fake news or deceptive posts for viral disinformation campaigns. To mitigate these risks, content authenticity scoring is employed using forensic media analysis, which assesses the credibility of content and cues human review when necessary. Visible digital watermarking is another vital strategy, marking synthetic media to indicate the need for verification. Additionally, screening publishers against known disreputable entities helps limit the reach of harmful actors.

Adversarial risks involve slight perturbations in input that can manipulate model behavior in dangerous ways. Anomaly detection systems are crucial for identifying out-of-distribution deviations, triggering further investigation. Rule-based blocking mechanisms are employed to shield against known triggers. Furthermore, adopting ensemble approaches renders adversarial attacks impractical, as they would require compromising multiple models.

Fraud risks arise from personalized phishing attempts that leverage generative text or image synthesis for mass social engineering. To counter this, allow lists are established to prevent unwanted messaging that lacks authenticated user consent. Intrinsic context watermarking techniques are utilized to bind outputs to their intended use case, thereby limiting misdirection. External partnerships also play a significant role in expanding perspectives on emerging threats.

While harmful outcomes from generative AI are often the result of human choices, implementing thoughtful governance is key. This involves balancing the freedom for creativity with responsible constraints, employing layered technical and relational policies to protect stakeholders. Such governance ensures that while the innovative potential of generative AI is harnessed, its risks are effectively managed to prevent misuse and harm.

7. Compliance Risks

Navigating the evolving landscape of global regulations on algorithmic systems, data privacy, and AI model requirements is crucial for maintaining legal and ethical compliance. This requires an agile approach to adapt to changing standards. The following sections outline key compliance risks and their mitigation strategies:

Data sovereignty confusion arises from conflicting national data sovereignty and data localization laws, leading to ambiguities in choosing permissible data storage locations while managing customers and users globally. To mitigate this, adopting multi-cloud hybrid storage solutions allows for the separation of datasets in compliance with distinct regulatory regimes. Employing federated learning keeps data partitioned locally, sharing only model learnings. Additionally, consulting with external policy experts is vital in navigating the complexities of overlapping jurisdictions.

The risk of incomplete feature documentation stems from insufficient tracking of data categories, feature engineering, and cohort coverage, which weakens support for external audits of ethical AI practices, especially with emerging certification standards. Mitigating this risk involves rigorous metadata persistence over dataset and pipeline stages, logging key facts for recall, including detailed data dictionaries. The creation of model cards provides snapshots that annotate intent, testing benchmarks, and advisories for stakeholders. Validating artifacts against external frameworks ensures mapping completeness.

Policy violations can occur when there is an over-reliance on manual inspection, potentially overlooking discrepancies between documented responsible AI principles and actual organizational practices. Automated policy conformance checks are essential for testing artifacts against codified standards, providing necessary guardrails. Internal audits are conducted to assess end-to-end alignment and address any identified gaps. Furthermore, external audits benchmark the substantive maturity of practices and policies.

Continuously assessing and adapting to the evolving regulatory and ethical AI policy obligations across jurisdictions is critical. Bridging the gaps between documentation, practices, and principles is vital for sustaining compliance, which is fundamental for maintaining customer trust and the operational license to innovate responsibly. This approach ensures that organizations not only adhere to current standards but are also well-positioned to respond to future regulatory changes and ethical considerations in the field of AI.

Fostering Responsible AI Through Partnerships and Systems Thinking

Adopting a multidimensional approach to assess and mitigate risks in critical areas is essential for fostering resilience and sustained innovation in AI applications, ensuring they are aligned with human values. This involves a blend of technical, procedural, and relational governance to create a holistic protection framework, enabling generative models to assist rather than impair.

However, achieving comprehensive risk coverage necessitates collaboration across the entire AI ecosystem. This includes AI service providers offering platform infrastructure, AI model developers, integration consultants deploying within business contexts, end-user organizations, regulators providing guidance, and the communities impacted by these technologies. No single entity can fully account for the complexities of real-world usage scenarios on its own.

Cloud vendors responsible for maintaining deployment environments need to work in tandem with methodology experts. These experts play a crucial role in flagging model recommendations that require heightened scrutiny, based on risk profiles assessed by researchers. Additionally, community panels are instrumental in providing qualitative human judgment on outputs, balancing multiple ethical considerations.

Extensive partnerships within the global AI ecosystem, coupled with a systems thinking approach that crosses traditional silos, are vital for gaining a comprehensive understanding. Such collaborations are key to strengthening guardrails, facilitating open communication channels for addressing concerns, and sustaining trust. This is crucial for the safe integration of AI across industries, ensuring equitable prosperity for all.

The commitment to an inclusive evaluation of generative AI risks is not just about accelerating innovation but also about preventing progress from being stalled by fragmented efforts. Organizations that benefit from AI technologies have a responsibility to contribute to multilateral efforts governing these technologies, ensuring their use is for the common welfare. This collaborative approach is essential for navigating the complexities of AI applications responsibly, ensuring they are used ethically and beneficially in various contexts.

Integrating AI Risk Mitigation with Enterprise Risk Management (ERM)

The integration of AI risk mitigation strategies within the broader framework of Enterprise Risk Management (ERM) is crucial for organizations leveraging AI technologies. This integration ensures that the unique risks associated with AI are systematically identified, assessed, and managed in alignment with the organization's overall risk management processes. The following discussion outlines key aspects of this integration:

1. Alignment with ERM Objectives: The first step in integrating AI risk mitigation with ERM is to align AI risk management objectives with the broader goals of ERM. This involves understanding how AI risks can impact the organization's strategic goals, financial stability, operational efficiency, and compliance obligations. By aligning AI risk mitigation with these objectives, organizations can ensure that their AI initiatives support rather than undermine their broader risk management strategy.

2. Risk Identification and Assessment: AI introduces specific risks, such as data privacy concerns, bias and fairness issues, and model explainability challenges. These risks need to be identified and assessed within the context of the organization's ERM framework. This involves conducting thorough risk assessments that consider the likelihood and potential impact of each risk, and how they might interplay with other enterprise risks.

3. Risk Response Integration: Once AI risks are identified and assessed, appropriate risk responses must be integrated into the ERM strategy. This includes developing mitigation strategies specific to AI, such as implementing data governance policies, ensuring transparency in AI decision-making, and establishing robust AI ethics guidelines. It's important that these AI-specific responses are harmonized with broader risk response strategies within the organization.

4. Continuous Monitoring and Reporting: AI systems are dynamic, and their risks can evolve rapidly. Therefore, continuous monitoring of AI risks is essential. This should be integrated into the organization's overall risk monitoring system. Regular reporting on AI risk status, including emerging risks and the effectiveness of mitigation strategies, should be part of the organization's risk reporting process.

5. Cross-Functional Collaboration: Effective integration of AI risk mitigation within ERM requires collaboration across various functions within the organization, including IT, data science teams, legal, compliance, and risk management departments. This ensures a holistic approach to AI risk management, with each department bringing its expertise to the table.

6. Training and Awareness: Building awareness and understanding of AI risks across the organization is essential. This involves training staff at all levels on the basics of AI risks, the importance of adherence to AI ethics and compliance standards, and their role in supporting effective AI risk management.

7. Adaptability and Learning: AI and its associated risks are continuously evolving. Therefore, the integration of AI risk mitigation with ERM should be adaptable and incorporate learning mechanisms. This includes staying updated with the latest developments in AI technology, regulatory changes, and best practices in AI risk management.

Integrating AI risk mitigation within an organization's ERM framework is essential for a comprehensive and cohesive approach to risk management. This integration ensures that AI is used responsibly and effectively, supporting the organization's goals while managing the unique risks it brings.

The Executive Order as a Turning Point in AI Regulation

In October 2023, a landmark Executive Order issued by President Biden marked a turning point in AI regulation, highlighting the transition of AI from a purely innovative technology to a subject of broader leadership responsibility across government and private sectors. This directive is set to shape the regulatory landscape of AI in the years ahead, underscoring its significance in various industries.

The principles and actions outlined in the order present both challenges and opportunities for organizations employing AI. It redistributes certain risks associated with algorithmic systems, mandating safety guardrails and enhanced transparency. This shift towards accountable governance models is poised to bolster consumer confidence and trust in AI applications.

As AI increasingly infiltrates critical sectors like healthcare and transportation infrastructure, the implications of irresponsible AI use become more consequential. The Executive Order catalyzes governmental efforts to address pressing issues such as bias, privacy breaches, and anticompetitive practices, signaling a definitive end to the era of laissez-faire AI experimentation.

Key Impacts of the Executive Order

The order requires safety testing and evaluations for powerful systems like large language models. Developers of dual-use AI must report training processes and model performance results to federal agencies. Technology executives should prepare for heightened security auditing around access to sensitive models.

While supporting innovation the order restricts anticompetitive activities from dominant platforms that could disadvantage consumers and businesses reliant on AI. It promotes law enforcement collaboration to address growing concerns like AI-enabled cyberattacks and intellectual property violations. Legal teams will need to monitor these emerging regulations.

For government agencies and contractors, new positions like chief AI officers will coordinate implementation of trustworthy AI practices. Across sectors, the order compels transparent communication on how AI systems impact the public through data gathering and decision-making. Both technologists and executive leadership must strategize around these rising expectations.

Preparing for a New Regulatory Paradigm

The Executive Order places a significant emphasis on developing technical guidelines and standards for reliable and safe AI usage. It entrusts scientific bodies like the National Institute for Standards and Technology with the task of formalizing best practices that will inform policymaking.

Private companies are encouraged to view ethical AI risks not as inevitable externalities but as integral priorities requiring internal governance review. By proactively aligning with forthcoming regulatory frameworks, organizations can develop the institutional knowledge necessary for long-term compliance.

Executive leaders should perceive new regulations not as punitive constraints but as opportunities for public-private collaboration. Active participation in shaping the AI safety agenda can result in regulations that balance rapid progress with necessary societal guardrails. The actions and decisions made by the industry today regarding responsible AI will significantly influence the direction of future legislative actions.

Navigating the Future with a Robust AI Risk Mitigation Framework

As we stand on the cusp of a new era in artificial intelligence, the importance of a comprehensive AI Risk Mitigation Framework cannot be overstated. This chapter has systematically explored the multifaceted risks associated with AI, encompassing output quality, data security, privacy, bias and fairness, transparency and explainability, misuse and harms, compliance, and the implications of the landmark Executive Order on AI regulation. Each section has underscored the necessity of proactive and strategic approaches to manage these risks effectively.

The overarching theme is clear: the path to harnessing the full potential of AI lies not only in technological advancement but equally in responsible stewardship. As AI continues to permeate every facet of our lives—from healthcare and transportation to education and entertainment—the need for robust risk mitigation strategies becomes increasingly paramount. These strategies must be dynamic, evolving in tandem with the rapid pace of AI development and the shifting landscapes of legal, ethical, and social norms.

The integration of AI risk mitigation within enterprise risk management highlights the need for a holistic approach that aligns AI initiatives with broader organizational goals. It's a call to action for cross-functional collaboration, continuous learning, and adaptation. The Executive Order on AI regulation serves as a pivotal moment, a clarion call for the industry and governments to work hand-in-hand to ensure that AI evolves in a manner that is safe, equitable, and beneficial for all.

In conclusion, this AI Risk Mitigation Framework offers a blueprint for organizations to navigate the complexities of AI deployment responsibly. It emphasizes the need for continuous assessment, the importance of ethical considerations, and the value of collaborative efforts to shape a future where AI is a force for good. As we embrace this journey, the decisions made today will undoubtedly shape the trajectory of AI for generations to come, making it imperative that these decisions are made with foresight, responsibility, and a deep commitment to the betterment of society.

Chapter 8

AI Adoption Roadmap

"To master a new technology, you have to play with it." — Jordan Peterson

A re you struggling to harness the full potential of Artificial Intelligence in your organization? You're not alone. While AI holds the promise of transformative power and competitive edge, the reality for most organizations is a maze of complexity and unfulfilled potential. This chapter is dedicated to unraveling this enigma through a well-structured strategic framework for AI adoption.

In the realm of digital transformation, AI is often seen as a silver bullet. Yet, the journey from ambition to actualization is fraught with challenges. This chapter demystifies this journey, offering a pragmatic, yet ambitious, strategic framework. Our comprehensive research across various industries has distilled the intricate process of AI adoption into a clear and actionable pathway.

This AI Adoption Framework, developed from an analysis of diverse AI implementations, is a beacon guiding organizations through the murky waters of AI integration. It acts as both a maturity model and a roadmap, aiding leaders in benchmarking their progress, fine-tuning their strategies, and systematically elevating their AI initiatives.

We've identified five critical phases in an organization's AI journey, each marking a significant milestone from initial experimentation to fully realized impact:

1. **Planning** - Laying the groundwork.

2. **Piloting** - Testing the waters of feasibility.

3. **Institutionalization** - Embedding AI competencies.

4. **Scaling** - Expanding to achieve widespread application.

5. **Differentiation** - Attaining transformative impact.

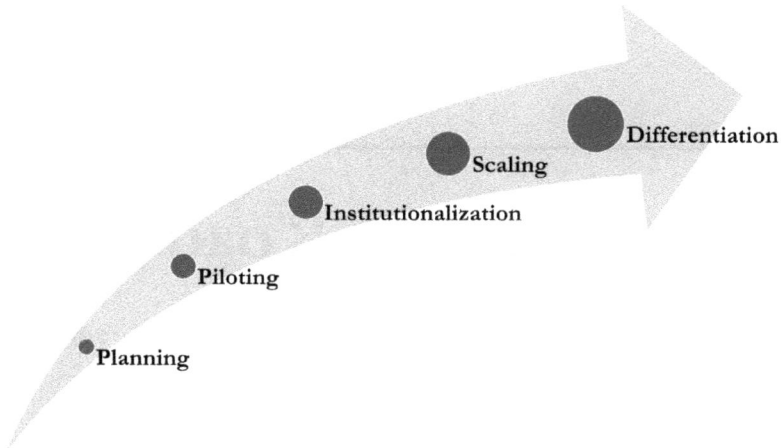

Figure 8-1: The Five Phases in the AI Adoption Roadmap

In these phases, we don't just outline objectives and tasks, but also provide success metrics and identify potential risks. Recognizing that misalignment with an organization's current maturity can derail progress, we emphasize the importance of strategic alignment with each phase.

This framework is not just a roadmap; it's a tool for leaders to:

- Assess their organization's current AI maturity and capabilities.

- Identify objectives that align with their current phase.

- Develop projects that are in sync with their AI maturity.

- Measure progress with phase-specific metrics.

- Navigate inherent risks, and recalibrate strategies to maintain momentum.

Ultimately, this framework isn't just about adopting AI; it's about doing it right. It's a blueprint for navigating the complex journey of AI integration, balancing innovation with responsibility. By following this framework, leaders can not only unleash the full potential of AI in their enterprises but also ensure a sustainable and ethical integration into their organizational fabric.

In the following sections, we delve deeper into each of these five phases, laying out a detailed map for your journey through the world of AI adoption.

1. Planning Phase: Laying the Groundwork for AI Success

The Planning phase is the critical foundation that sets up AI adoption for eventual success. Many organizations underestimate the importance of this phase and rush ahead to piloting without proper planning and preparation. However, inadequate planning leads to false starts, wasted resources, and initiative failures down the line. As the adage aptly warns, "Failing to plan is planning to fail."

Key Elements of the Planning Phase

1. **Use Case Identification**: The heart of the Planning phase is identifying 3-5 promising AI use cases that align with your business objectives and offer measurable value. This requires a collaborative approach, bringing together cross-functional teams of business, data, and AI experts. These workshops should culminate in clearly defined use cases, each championed by an executive sponsor ready to support the pilot phase both financially and strategically.

2. **Foundation Development**: A successful AI initiative needs a strong foundation. This involves addressing skill gaps with targeted reskilling programs, laying down infrastructure requirements, estimating budgets, and establishing governance frameworks. This stage involves formulating AI policies that cover risk management, ethical considerations, and regulatory compliance.

3. **Executive Buy-In**: Securing executive support is a critical milestone in the Planning phase. It involves demonstrating the potential of AI initiatives and substantiating the business case for each use case. Building a consensus among leadership for AI investments and sponsorship is essential to move forward confidently.

4. **Knowledge Development**: Building AI expertise within the core team is a key focus. This involves a deep dive into AI technologies, methodologies, and best practices, often supplemented by insights from external experts. This stage is about preparing the team not just with technical knowledge but also with an understanding of the broader AI landscape.

Success Metrics for the Planning Phase

- Percent of identified use cases with assigned executive sponsors

- Completion rate of skills assessments for the AI team

- Development of risk governance policies and AI ethics principles

- Favorable risk-benefit analysis for top use cases

- Percent of employee AI literacy for priority roles

- Number of data sources assessed for viability per use case

- High-level frameworks created to guide use case prototypes

- Degree of consensus on priority use cases among stakeholders

Achieving these metrics effectively prepares the initiative for a successful transition to the piloting phase.

Key Risks in the Planning Phase

- Insufficient executive sponsorship and budget commitments

- Resource/talent gaps regarding AI and analytics skills

- Lack of quality, integrated data across use cases

- Unable to clearly determine use case viability and priority

- Poor risk anticipation regarding impacts

- Overlooking important ethical considerations

- Unrealistic expectations on AI capabilities and timelines

- Insufficient coordination across involved teams

Mitigating these risks upfront prevents false starts mid-stream and significantly improves success odds in later AI adoption stages once piloting and scaling commences. Investing rigor upfront determines how well foundations stand the test of time.

In essence, the Planning phase is about diligence and foresight. Neglecting this stage can lead to complexities later, which are much harder to resolve midstream. Thus, a thorough and meticulous approach in the Planning phase is not just beneficial but necessary for a successful and sustainable AI adoption journey.

2. Pilot Phase: Testing AI in Real-Time

Following a robust planning stage, organizations embark on the Pilot phase, a critical juncture where AI's potential is tested through practical applications. This phase is about transforming high-potential use cases into tangible prototypes and validating key hypotheses.

Focus Areas of the Pilot Phase

1. **Infrastructure Setup:** The pilot phase creates the initial infrastructure for AI development, operations, and monitoring. This includes data pipelines, model development environments, MLOps automation, testing harnesses, and other foundations to progress pilots and gather feedback.

2. **Execution & Measurement:** Teams progress through iterative build-measure-learn cycles to pilot AI solutions and evaluate outcomes. The emphasis lies more on validating value vs. progressing capabilities. Teams design pilot performance dashboards covering business impact metrics, model accuracy KPIs, technical metrics, and stakeholder feedback.

3. **Lesson Assimilation:** An underlying objective involves assimilating lessons from piloting to refine the broader AI strategy. What worked and what failed? Which use cases show more traction? Should adoption pace accelerate or decelerate based on pilot results? Teams codify insights to shape the path forward regarding tooling, skills, data, governance, and scaled delivery.

Success Metrics for the Pilot Phase

- Meeting or exceeding predefined KPIs related to business improvement.

- Achieving model accuracy, precision, and recall above set thresholds.

- Reducing turnaround times for key business processes.

- Garnering high satisfaction levels from pilot user groups.

- Over 80% of prototypes fulfilling specified requirements.

- Effective documentation of insights from iterative cycles.

- Positive outcomes from scaled pilot expansions.

- Low technical debt with reusable, portable prototype components

Achieving these metrics is indicative of the AI system's potential to add value to business processes and warrants further investment in its development.

Key Risks in the Pilot Phase

- Inability to fully simulate real-world conditions

- Poor data quality hindering model reproducibility

- Over customization leading to non-reusable components

- Underestimating infrastructure needs at scale

- Not designing with operationalization considerations

- Inadequate governance controls and safeguards

- Lack of monitoring once deployed to end-users

- Insufficient documentation for handoffs

Proactively addressing these risks is essential for ensuring that successful pilots can smoothly transition into integrated AI applications.

The Pilot phase serves as a vital checkpoint in the AI adoption journey. Successfully piloting AI applications not only demonstrates the technology's potential but also secures greater organizational support for further AI initiatives. Conversely, setbacks in this phase can dampen confidence and drain resources. A well-structured approach to piloting, coupled with diligent measurement and learning, sets the stage for the subsequent phases of institutionalizing, scaling, and differentiating AI capabilities.

3. Institutionalization Phase: Embedding AI Deep into the Organizational Fabric

The Institutionalization phase marks a pivotal transition from successful AI pilots to embedding AI capabilities deeply within the organization. This stage is about operationalizing AI, controlling risks, and preparing for responsible scaling.

Key Aspects of the Institutionalization Phase

1. **Center of Excellence (CoE)**: This phase sees the establishment of a CoE, a cross-disciplinary entity that becomes the hub for AI expertise, best practices, and support. Its role is to facilitate AI adoption across business units, evolving eventually into a self-service portal or platform. The CoE serves as a centralized resource for skills, platforms, and tools, streamlining AI development and governance.

2. **Governance Enablement**: Strengthening governance frameworks is essential before AI is scaled up. This involves formalizing model risk frameworks, establishing approval processes, and setting up monitoring practices. The goal is to ensure regulatory compliance, safety, quality, and responsible AI deployment. Consistency is reinforced through instructional materials and assessments.

3. **Platform Consolidation**: A key move in this phase is consolidating the diverse tools and infrastructures used during pilots into standardized, governed platforms. This involves creating interoperable pipelines that integrate data management, model development, MLOps, testing, and monitoring infrastructure.

4. **Process Definition**: The CoE leads the charge in defining comprehensive processes for AI development, validation, deployment, and monitoring, aligning them with DevOps principles. This includes detailed process maps for model requests, risk reviews, development sequences, handoffs, and stage approvals, ensuring rapid and reliable AI delivery with built-in quality.

Success Metrics for the Institutionalization Phase

- Reduced costs in coordinating AI projects

- Shorter time-to-value for model development

- Higher rates of model and pipeline reuse

- Lower instances of deviations from established governance protocols

- Reduced duplicate tools and environments

- Improved model handoff efficiency

- Increased autonomy for business users through self-service features

- Enhanced model performance through continuous learning

Achieving these signifies the increased maturity required for AI to reliably impact operations.

Key Risks in the Institutionalization Phase

- Platform inconsistencies leading to fragmentation

- Lack of integrated and automated governance mechanisms

- Business preferring custom solutions over standardized reusables

- Challenges in standardization due to diverse organizational needs

- Poor documentation affecting continuity and knowledge transfer

- Loosely defined processes leading to sprawl

- CoE's skillset limitations in supporting a wide range of needs

- Compliance gaps due to decentralized execution

Proactive mitigation of these risks is crucial for laying a solid foundation for scaling AI across the organization.

The Institutionalization phase is about turning ad hoc AI initiatives into an integrated part of the organizational strategy. It focuses on building competencies and establishing robust processes and governance structures. This foundation is essential for scaling AI effectively and responsibly across the enterprise, ensuring that AI systems are delivered reliably and in alignment with business goals.

4. Scaling Phase: Amplifying AI's Impact Across the Organization

After successfully embedding AI in the organizational framework, the Scaling phase is about expanding its reach and impact. This phase is characterized by increasing AI's breadth and depth across the enterprise, combining industrialization, democratization, infrastructure enhancement, and automation.

Key Aspects of the Scaling Phase

1. **Industrialization**: This aspect focuses on replicating established AI competencies across various business units, fostering decentralized AI adoption. The COE plays a crucial role here, providing platforms, tools, and expertise, thus enabling teams to develop contextual solutions efficiently. The emergence of workgroup data science augments centralized delivery.

2. **Democratization**: The aim is to make AI more intuitive and accessible, especially for business teams without specialized skills. This involves implementing no-code tools, AI-enhanced app stores, chatbots, and agent advisors. Emphasis is placed on user experience, making AI tools more user-friendly and widely accessible.

3. **Infrastructure Scaling**: To accommodate the growing number of AI projects and their complexity, infrastructure, platforms, and tooling are enhanced. This includes scaling up on-demand compute capabilities, version control systems, and reusable component libraries. These upgrades are designed to empower AI engineers to deliver more rapidly and without technical limitations.

4. **Automation Advancement**: By refining MLOps, DevOps, DataOps, and ModelOps practices, organizations aim to seamlessly integrate human-and-machine collaboration. Automation is incrementally applied throughout the machine learning lifecycle, from data collection to model deployment, thereby enhancing both speed and reliability.

Success Metrics for the Scaling Phase

- Increased AI adoption across business units

- Higher AI contribution to key innovation metrics

- Reduced timeframes for AI model development and deployment

- Lower compute costs per AI workload

- Greater suite of no-code AI tools and templates

- Increased user autonomy and productivity through AI tools

- Higher AI integration into products and services

- Greater interoperability between AI components

Meeting these metrics indicates the necessary expansion of AI capabilities to transform business operations.

Key Risks in the Scaling Phase

- Business units deviating from institutionalized practices

- Automation and tooling gaps hindering democratization

- Lack of a deliberate scaling approach and governance

- Inadequate support for increasingly complex workloads

- Self-service capabilities lagging behind demand

- Technical debt and custom fragmentation re-emerging

- Compliance gaps due to decentralized control

- Mission creep from lack of continuous alignment

Proactively addressing these risks ensures responsible and controlled expansion of AI capabilities, thereby facilitating uninterrupted progress in both productivity and innovation.

The Scaling phase is pivotal in transitioning AI from a functional tool to a transformative force. It centers on leveraging AI to not only augment existing processes but also to forge new pathways for innovation and strategic adaptability. This dual emphasis on both widespread implementation and efficiency optimization is crucial to averting stagnation in AI adoption. By scaling AI, organizations can revitalize their current operations and simultaneously liberate resources to explore novel solutions and adaptable strategies. Achieving this intricate balance of widespread AI tooling, advanced automation, and accessible technology is key to unlocking unique opportunities for differentiation and growth.

5. Differentiation Phase: Elevating AI to a Strategic Imperative

he Differentiation phase symbolizes the zenith of AI maturity, where its strategic integration not only enhances processes but also defines the competitive landscape. In this phase, AI becomes so seamlessly integrated into the organizational fabric that it propels breakthrough innovations, operational efficiency, and unparalleled customer experiences.

Key Aspects of the Differentiation Phase

1. **Strategic Infusion**: AI transcends its traditional roles to shape executive decision-making and innovate business models, products, and partnerships. It becomes a tool for redefining value propositions and creating unique offerings and experiences, moving beyond just optimizing costs and incremental improvements.

2. **Operational Integration**: AI systems become deeply embedded in the operational workflows, operating autonomously and learning in real-time. They evolve into self-sufficient entities, significantly enhancing human productivity by taking over predictable tasks and optimizing processes through continuous data analysis.

3. **Culture Cultivation**: Building an AI-first mindset becomes a leadership imperative. This involves a comprehensive change management strategy to realign roles, structures, and incentives to foster human-AI synergy. It's about embedding ethical AI principles into the organization's culture and removing barriers to AI adoption.

4. **Co-Innovation**: This stage encourages collaborative innovation, leveraging the collective strengths of internal AI teams, business units, and external partners. It's about exploring new avenues, from partnerships and

M&A activities to investments in startups, to drive major advancements and novel solutions.

Success Metrics for the Differentiation Phase

- Enhanced revenue and market share due to AI-driven strategies

- Increased profit margins through AI-optimized operations

- Groundbreaking innovations and new intellectual property creation facilitated by AI

- Accelerated time-to-market for AI-enabled products and services

- Improved customer retention and acquisition, supported by AI-augmented experiences

- Cumulative operational efficiency gains from integrated AI solutions

- Elevated employee productivity, creativity, and satisfaction from AI integration

- Comprehensive adherence to responsible AI principles across all solutions

Achieving these milestones signifies not only the realization of AI's value at scale but also a distinct competitive edge.

Key Risks in the Differentiation Phase

- The risk of commoditization from focusing solely on incremental AI applications

- A gradual dilution of responsible AI principles

- Diminished creative thinking and overreliance on automated decision-making

- Workforce skills atrophy as AI assumes more responsibilities

- Challenges in maintaining differentiation due to talent shortages

- Emerging cyber risks as operations increasingly depend on AI

- Reduced human oversight in AI systems due to feedback loop limitations

- Rapid shifts in market dynamics that can reduce the relevance of current AI strategies

Proactively mitigating these risks is crucial. It involves fostering a symbiotic relationship between human creativity and abilities and the capacities of automated AI systems, ensuring sustainable differentiation.

In conclusion, at this pinnacle stage, we witness the emergence of an AI-native enterprise. In this state, AI becomes so intricately integrated into the enterprise that it becomes virtually invisible, functioning seamlessly within the organization's fabric. The core capabilities of such an enterprise reach a point where automation not only becomes self-sustaining but also self-enhancing, evolving through recursive improvements. To sustain leadership in this advanced stage, it is imperative to continuously drive next-generation innovations. By strategically, operationally, and culturally embedding AI into the core of business practices, a company not only establishes a significant competitive edge in the current market but also paves the way for the advent of future AI-native enterprises, setting new benchmarks for success in the industry.

Climbing the Summit: Navigating the Path to AI Actualization

The AI Adoption Framework presents a comprehensive and strategic roadmap designed to unlock the vast potential of AI within organizations. This blueprint skillfully balances ambitious goals with pragmatic approaches, offering tailored guidance that evolves alongside an organization's growing AI capabilities. The journey delineated by this framework is a transformative progression from initial, often chaotic experimentation to a state of embedded AI mastery. By adhering to the framework's carefully structured guidance, aligned with each stage of maturity, organizations can expect to see compounding benefits as AI becomes increasingly integrated across all facets of the enterprise.

Central to this journey is leadership commitment, which must synergize with the framework's insightful recommendations and phase-specific strategies to drive AI success. Following this framework diligently sets in motion a dynamic process of change management, transitioning organizations from sporadic AI experimentation to becoming AI-empowered and ultimately AI-native entities. In this evolutionary process, AI transitions from isolated, narrow applications to broader optimizations. Over time, it revolutionizes core operations, catalyzing innovation in products and services.

As AI becomes intricately woven into the strategic fabric of an organization, it fosters a profound symbiosis between human capabilities and machine intelligence. This harmonious collaboration yields sustainable value, balancing technological benefits with enhancements in skills, governance, and responsible utilization throughout the journey. Ultimately, competitive advantage accrues to those leaders who thoughtfully and responsibly integrate AI into their organizational DNA.

Realizing the Vision of an AI-Native Enterprise

The strategic framework for AI adoption acts as both a compass and a safeguard, guiding organizations on their journey to transform AI's potential into a tangible and enduring impact. It stresses the crucial need for aligning AI deployment with thorough maturity assessments, ensuring that each phase of adoption builds logically upon the previous one. The activities and strategies within this framework are designed to seamlessly transition organizations to successive phases, ensuring a holistic and well-coordinated integration of AI.

By diligently and strategically implementing this framework, organizations can expect AI to become a pervasive and almost indiscernible presence, intricately woven into the very fabric of their systems and processes. This is the defining characteristic of an AI-native enterprise. In such an enterprise, AI transcends its role as a mere tool or solution; it becomes a fundamental aspect of the organization's operational, innovative, and competitive landscape. This framework is not merely a route to utilizing AI technology; it is a comprehensive blueprint for creating a future where AI is central to business evolution and success, driving transformation not just in technology, but in every facet of the organization.

Chapter 9

AI-Powered Business Transformation Playbook

"Transformation literally means going beyond your form." — Wayne Dyer

A re you prepared to navigate the seismic shifts in today's business landscape? How can your organization not only adapt but thrive in the age of artificial intelligence? In this crucial chapter of the 'AI-Powered Business Transformation Playbook,' we address these pressing questions, guiding business leaders through the labyrinth of AI-driven enterprise metamorphosis.

Imagine a future where business is not just enhanced but fundamentally transformed by AI. This journey begins with identifying and tackling the challenges of today's market. We commence with business augmentation, navigating through intricate pathways of business transformation, and ultimately arriving at a complete business reinvention. These phases are scaffolded by esteemed research frameworks like MIT's Digital Maturity Model and McKinsey's Digital Quotient, honing in on operational excellence, customer intelligence, and ecosystem leverage.

Our focus here extends beyond the mere adoption of technology. We explore the unparalleled prospects ushered in by emergent Generative AI. When embedded into an organization's ethos, AI's boundless generative capabilities promise to unlock futures we are yet to envision, paving the way for novel markets, unique experiences, and innovative value chains.

For business leaders, the journey is clear but challenging: evolve from achieving pinpoint efficiencies to embracing enterprise-wide integration, and ultimately, spearheading market reinvention. This path mirrors the maturation of AI technologies, evolving from basic analytics to deep learning and sophisticated generative models.

This chapter ignites a conversation about the synergy between progressively advanced AI algorithms and human ingenuity, fostering a perpetual cycle of co-creation. AI expands our challenge horizon while simultaneously enriching our solu-tion toolkit. We envision a future of AI-human symbiosis, continuously redefining the frontiers of innovation, personalization, and category creation.

While the previous chapter outlined a five-phase roadmap for AI adoption, here we shift our gaze to the business realm, emphasizing the leader's perspective in leveraging AI for organizational transformation. This is more than a technological upgrade; it's a strategic leap into redefining competitive differentiation and securing market leadership in the AI era.

Use this guidebook not just as a map but as a compass, orienting your organization's vision and strategy across these transformative phases. Mastering this journey is about transcending technology to propel your business to new, uncharted heights.

Figure 9-1: The Three Stages of AI-Powered Business Transformation

Stage 1: Business Augmentation - Unleashing AI's Potential in Key Business Areas

The first stage, 'Business Augmentation,' guided by Forrester's Digital Process Op-timization model, is where artificial intelligence begins to redefine the business landscape. In this phase, AI is strategically integrated into specific areas of the business to enhance efficiency, decision-making, customer experiences, and risk management.

Key Objectives

- **Streamlining Workflows**: Here, AI is used to automate repetitive tasks across operations. This includes data entry, document processing, and other routine activities, employing tools like robotic process automation and AI algorithms. The goal is to transform these mundane tasks into efficient, automated processes.

- **Enhancing Decision-Making**: AI's role in decision-making is pivotal. By building predictive analytics models, businesses can forecast and strategize resource planning, optimize processes, and conduct more accurate risk analysis. This is not just about processing data but about deriving actionable insights to make informed decisions.

- **Strengthening Customer Experiences**: Utilizing AI technologies like chatbots, recommendation engines, and personalization algorithms, businesses can tailor their customer engagement strategies. This leads to more personalized, responsive, and engaging customer interactions.

- **Mitigating Risks**: AI systems play a crucial role in identifying and managing risks. Through advanced analytics, businesses can detect anomalies, predict maintenance needs, and conduct fraud analysis, significantly reducing potential threats and vulnerabilities.

Strategic Moves

- **Pursuing High ROI Initiatives**: Identifying and implementing high ROI proof-of-concept projects are essential. These projects should demonstrate tangible outcomes and have commercial viability, serving as exemplars of AI's potential in the business.

- **Building Data Infrastructure**: A robust data infrastructure is foundational. This involves curating annotated datasets, establishing efficient data pipelines, and adhering to governance protocols, ensuring the data's quality and security.

- **Fostering AI Talent and Leadership**: Developing a skilled internal team is vital. This includes not just technical experts like data scientists and engineers but also domain experts who understand both the technology and its business applications. Additionally, educating leadership on AI's capabilities and guiding them through the organizational changes it entails is crucial for a cohesive transformation journey.

Measuring Success

- **Operational Efficiencies**: Success in this phase is marked by measurable improvements in operational processes, accuracy, and compliance, as automation takes center stage.

- **Cost Reduction**: Significant cost savings are expected through streamlined workflows and preventive analytics, reducing the need for extensive manual labor and resource allocation.

- **Revenue Growth**: Enhanced customer experiences and retention strategies, powered by AI, contribute to revenue growth, indicating the successful application of AI in customer-facing roles.

Examples of Stage 1

1. Automating Customer Service with AI

- **Context**: A telecommunications company faces challenges with high call volumes and customer service quality.

- **AI Implementation**: The company integrates AI-powered chatbots to handle routine customer queries. These bots use natural language processing to understand and respond to customer needs, providing quick and accurate information.

- **Impact**: Call wait times reduce significantly, customer satisfaction improves, and the human customer service team can focus on more complex issues.

2. Enhancing Supply Chain Efficiency

- **Context**: A global retailer struggles with inventory management across its diverse range of products and locations.

- **AI Implementation**: The retailer deploys AI algorithms for predictive analytics, forecasting demand based on factors like market trends, historical sales data, and seasonal variations.

- **Impact**: Inventory management becomes more precise, reducing overstock and stockouts, leading to cost savings and improved availability of products for customers.

3. Streamlining Financial Operations

- **Context**: A financial services firm grapples with the time-consuming process of loan approvals and risk assessments.

- **AI Implementation**: AI models are introduced to analyze loan applications, using historical data to assess credit risk and predict loan defaults.

- **Impact**: The loan approval process becomes faster and more accurate, enhancing customer experience and reducing the risk of bad debts for the firm.

4. Personalizing Marketing Efforts

- **Context**: An e-commerce platform seeks to improve customer engagement and sales conversions.

- **AI Implementation**: The platform uses machine learning to analyze customer browsing and purchase history, creating personalized product recommendations for each user.

- **Impact**: Customers receive more relevant product suggestions, leading to higher engagement rates and increased sales.

5. Improving Healthcare Diagnostics

- **Context**: A healthcare provider wants to enhance the accuracy and speed of medical diagnostics.

- **AI Implementation**: AI-powered image recognition tools are used to analyze medical images like X-rays and MRIs, assisting doctors in identifying anomalies quickly.

- **Impact**: Diagnostics are more accurate and faster, leading to better patient outcomes and more efficient use of medical resources.

6. Optimizing Manufacturing Processes

- **Context**: A manufacturing company faces challenges in maintaining equipment and minimizing downtime.

- **AI Implementation**: Predictive maintenance algorithms are utilized to monitor equipment health in real-time, predicting failures before they occur.

- **Impact**: Unexpected downtime reduces significantly, production efficiency improves, and maintenance costs decrease.

These examples illustrate how businesses in Stage 1 leverage AI to optimize existing processes, enhance decision-making, and improve customer experiences, setting a solid foundation for further AI integration and transformation.

In essence, the Business Augmentation stage is where AI begins to show its transformative power, optimizing business operations while laying the groundwork for the extensive journey of AI integration ahead.

Stage 2: Business Transformation - Embracing AI to Reshape Business Models

Entering the second stage, 'Business Transformation,' companies leverage the foundational AI capabilities established in the augmentation phase to fundamentally reshape their business models. This phase is about hybrid transformation, integrating AI to innovate products, expand market reach, build ecosystems, and achieve hyper-personalization.

Key Objectives

- **Launching Data-Rich Products**: In this era of data-driven innovation, embedding predictive analytics, personalization algorithms, and automation capabilities into new products is key. These data-rich products are not just smart; they are intuitive, adapting to user needs and preferences.

- **Expanding to Adjacent Spaces**: Armed with deep customer and market insights, companies can now strategically venture into adjacent and untapped market spaces. This expansion is fueled by the intelligence gathered through AI, allowing businesses to identify and capitalize on new opportunities.

- **Architecting Platform Ecosystems**: This involves creating symbiotic partnerships and networks, where third-party developers and data interchange play pivotal roles. Powered by AI, these ecosystems thrive on collaboration and shared innovation.

- **Advancing Hyper-Personalization**: By connecting previously siloed datasets, businesses can craft a unified view of customer intelligence. This enables granular personalization at an unprecedented scale, catering to individual customer preferences across various touchpoints.

Strategic Moves

- **Focused Business Model Bets**: Companies should strategically invest in areas with high growth potential that align with their competitive strengths. These selective investments are the catalysts for outsized growth.

- **Doubling Down on Differentiation**: Unique offerings, experiences, and partnerships, driven by a company's unique assets and data, set it apart in the market. This differentiation is crucial in an AI-driven business landscape.

- **Operationalizing Data Network Effects**: Breaking down data silos and enabling real-time enterprise-wide data access creates a powerful network effect. Building feedback loops between systems enhances this effect, driving continuous improvement.

- **Nurturing an 'AI-First' Culture**: Encouraging a culture of cross-functional collaboration, continuous experimentation, and feedback centered around AI capabilities is fundamental. This culture shift ensures that AI is at the heart of every business decision and innovation.

Measuring Success

- **AI-Based Product Performance**: Assessing the success of offerings integrated with AI, such as prediction engines and conversational interfaces, is key to understanding their market impact.

- **Market Expansion Evaluation**: Monitoring the pace and effectiveness of using AI-driven intelligence to enter new segments, geographies, and categories provides insight into the company's adaptive growth.

- **Ecosystem Leverage Quantification**: Measuring the growth of the developer network, ROI from partnerships, and opportunities from data interchange reflects the success of the ecosystem strategy.

- **Customer Journey Mapping**: The precision with which unified data captures complete omnichannel customer behavior is a crucial metric, showcasing the effectiveness of hyper-personalization strategies.

Examples of Stage 2

1. Creating AI-Driven Fashion Retail Platforms

- **Context**: A fashion retail company seeks to revolutionize its online shopping experience.

- **AI Implementation**: The company develops an AI-driven platform that offers virtual try-on features and style recommendations based on customer preferences and body types.

- **Impact**: Customers enjoy a highly personalized and interactive shopping experience, leading to increased customer loyalty and sales.

2. Developing Smart Healthcare Solutions

- **Context**: A healthcare organization aims to enhance patient care through technology.

- **AI Implementation**: The organization introduces AI-driven telehealth services and remote monitoring systems that provide real-time patient data and predictive health insights.

- **Impact**: Patients receive personalized and proactive healthcare, improving treatment outcomes and expanding the organization's reach to remote areas.

3. Innovating in the Automotive Industry

- **Context**: An automotive manufacturer wants to stay ahead in the competitive market.

- **AI Implementation**: The company starts producing AI-integrated smart cars with advanced features like autonomous driving, predictive maintenance, and personalized in-car experiences.

- **Impact**: The brand establishes itself as a leader in innovation, attracting tech-savvy customers and setting new standards in the automotive industry.

4. Transforming Real Estate with AI Analytics

- **Context**: A real estate firm looks to optimize property management and investment strategies.

- **AI Implementation**: The firm uses AI to analyze market trends, property valuations, and investment risks, offering clients data-driven insights for real estate decisions.

- **Impact**: Clients make more informed investment choices, and the firm gains a reputation for delivering value-added services, enhancing its market position.

5. AI-Powered Environmental Conservation

- **Context**: An environmental organization focuses on wildlife conservation and habitat protection.

- **AI Implementation**: The organization employs AI algorithms to analyze satellite imagery and sensor data for monitoring wildlife populations and detecting illegal activities.

- **Impact**: Conservation efforts become more effective and efficient, leading to better protection of endangered species and habitats.

6. Revolutionizing Media Consumption with AI

- **Context**: A media company wants to enhance content delivery and user engagement.

- **AI Implementation**: The company introduces an AI-driven platform that curates personalized content for users based on their viewing history and preferences.

- **Impact**: User engagement skyrockets, and the platform sees increased viewership and advertising revenue.

These examples demonstrate how businesses in Stage 2 of the AI journey leverage AI to not just improve but transform their business models. By integrating AI into their core operations and offerings, they unlock new potentials, expand their market reach, and establish new standards in their respective industries.

In the Business Transformation phase, the focus is on harnessing AI's power to not just enhance but revolutionize business ecosystems, unlock deep intelligence, and foster innovation at the nexus of data and differentiation.

Stage 3: Business Reinvention - Pioneering the Future with Generative AI

In the final stage, 'Business Reinvention,' organizations, having achieved digital maturity, are encouraged by MIT's insights to completely reimagine their value propositions and industries. This stage is about boldly harnessing Generative AI to innovate and redefine business models, products, and experiences.

Moonshot Objectives

- **Incubating Disruptive Business Models**: This involves inventing scalable platforms, products, and services with AI embedded at their core. These are not just improvements but radical innovations that redefine industries.

- **Launching Smart Products 2.0**: Embedding ambient intelligence into physical-digital offerings, these next-generation products bring together automation and predictive personalization for enhanced user experiences.

- **Exploring Sci-Fi Scenarios**: Engaging in speculative design, companies brainstorm futuristic scenarios of human-AI collaboration. This is about pushing the boundaries of imagination with customers, partners, and vendors.

- **Manifesting Ambient Experiences**: Utilizing technologies like biometrics, computer vision, and AR/VR, businesses can create contextually intelligent omni-channel experiences. This is about making technology not just smart but seamlessly integrated into the user's environment.

Bold Strategic Bets for Sustained Innovation

- **Exploring Commercial Viability of Moonshots**: Visionary ideas are grounded in reality through rigorous viability, monetization, and adoption feasibility analyses. It's about balancing big ideas with practical execution.

- **Architecting for Perpetual Adaptability**: Designing technology architectures that are modular and dynamically updatable allows businesses to evolve in real-time as new breakthroughs emerge, ensuring perpetual relevance.

- **Harnessing Extreme Personalization**: By expanding real-time data networks, businesses can continuously optimize and customize interactions and content, offering a level of personalization that is not just tailored but predictive and intuitive.

- **Expanding the Frontiers of Imagination**: Companies are encouraged to explore uncharted territories, uncover hidden needs, and stretch possibilities through AI synthesis, constantly pushing the limits of what's conceivable.

Measuring Transformational Impact

- **Evaluating Exponential Traction**: Success is measured by the pace of user growth, revenue acceleration, and valuation expansion of AI-based business models. It's about capturing the market at an exponential rate.

- **Measuring Category Domination**: Analyzing shifts in market share towards AI-powered products or services provides a clear indicator of industry leadership and influence.

- **Quantifying Ecosystem Leverage**: The success of building synergistic partnerships centered around enterprise assets is assessed by their network effects and value addition.

- **Tracking Brand Equity Velocity**: Estimating the brand's value in terms of utility, affinity, and customer delight highlights the impact of providing intuitively delightful experiences.

Examples of Stage 3

1. Revolutionizing Urban Planning with AI

- **Context**: A city planning and development firm seeks to create smarter, more sustainable urban environments.

- **AI Implementation**: The firm utilizes Generative AI to simulate and design urban layouts, incorporating factors like traffic flow, environmental impact, and community needs.

- **Impact**: The result is a generation of innovative, sustainable city designs that optimize living conditions, reduce environmental footprints, and promote efficient resource use.

2. Disrupting the Entertainment Industry

- **Context**: A production company wants to redefine storytelling and viewer engagement.

- **AI Implementation**: By using Generative AI, the company creates interactive, adaptive storylines where content evolves based on viewer choices and reactions, offering a personalized entertainment experience.

- **Impact**: This reinvention transforms passive viewing into an interactive experience, setting a new standard in personalized entertainment and audience engagement.

3. Pioneering AI in Fashion Design

- **Context**: A fashion brand aims to lead in innovation and creativity.

- **AI Implementation**: The brand employs Generative AI to create unique, trendsetting designs based on global fashion trends, material innovations, and consumer preferences.

- **Impact**: The brand stands out for its cutting-edge, AI-generated collections, attracting a new generation of fashion enthusiasts and redefining industry standards.

4. Transforming Healthcare with Predictive Medicine

- **Context**: A healthcare provider strives to offer predictive and preventive care.

- **AI Implementation**: Leveraging Generative AI, the provider develops models that predict individual health risks and offer personalized preventive care plans.

- **Impact**: Healthcare becomes more proactive and personalized, significantly improving patient outcomes and paving the way for a new era in healthcare services.

5. Advancing Agricultural Practices

- **Context**: An agribusiness company focuses on sustainable and efficient farming methods.

- **AI Implementation**: The company uses Generative AI to analyze and predict optimal planting patterns, irrigation schedules, and crop rotations, considering environmental conditions and climate change impacts.

- **Impact**: The adoption of these AI-driven practices leads to higher crop yields, reduced environmental impact, and sets new sustainability benchmarks in agriculture.

6. Redefining Customer Service with AI Agents

- **Context**: A multinational corporation seeks to enhance global customer support.

- **AI Implementation**: The corporation develops advanced AI agents capable of understanding and solving complex customer issues, providing support in multiple languages, and learning from each interaction.

- **Impact**: Customer service is revolutionized with 24/7 support, high problem-solving efficiency, and a personalized customer experience, significantly boosting customer satisfaction and loyalty.

These examples illustrate how businesses in Stage 3 embrace Generative AI to not only innovate within their sectors but to completely redefine them. By harnessing the power of AI to create groundbreaking solutions and experiences, these companies pioneer new frontiers and establish themselves as leaders in the age of AI-driven business reinvention.

In this phase, 'Business Reinvention', generative imagination is both the fuel and the beacon, guiding companies to conceive and realize groundbreaking business models, offerings, and interfaces. This stage is characterized by a symbiotic collaboration between humans and AI, reshaping industries on a global scale.

Metamorphosis to an AI-Native Enterprise

What does it mean to be an AI-native enterprise in today's rapidly evolving business landscape? How can organizations leverage the full spectrum of AI's capabilities to not only stay relevant but lead the charge in innovation? This playbook serves as a strategic compass, guiding enterprises through a transformative journey across three escalating stages of AI integration.

- **Stage 1 – Questioning the Status Quo**: How can AI refine and enhance our current operations? This stage is about taking the first bold steps towards change. Imagine an enterprise where routine tasks are automated, decision-making is data-driven, and customer experiences are personalized – all powered by AI.

- **Stage 2 – Exploring New Possibilities**: How can AI take us beyond our current boundaries? Here, businesses start to expand and transform their core operations. Think of a healthcare provider using AI to develop new diagnostic tools, or a manufacturer leveraging AI to enter the realm of smart, connected products.

- **Stage 3 – Envisioning the Unimaginable**: What uncharted territories can we explore with AI? This stage is about reimagining and reinventing business models. Consider a world where AI helps create sustainable cities or where AI-driven platforms connect global talents in ways never thought possible.

This journey represents an intellectual evolution, challenging enterprises to progressively broaden their vision and ambition with AI. It's a cyclical process, with each iteration pushing the innovation envelope further.

The ultimate goal is to establish a symbiotic relationship where human ingenuity and AI's capabilities amplify each other. This positive feedback loop does not just redefine business frontiers; it ensures that enterprises embracing this cycle maintain a continually evolving competitive edge.

Embarking on this AI-Native journey requires courage to venture into uncharted territories. The rewards, however, are immense – creating new competitive paradigms unrestrained by traditional boundaries. In the current business landscape, integrating AI is no longer a strategic option but an existential necessity.

To thrive in the AI era, enterprises must transform into natively intelligent entities, continuously primed for innovation and market leadership. The future beckons for those prepared to answer its call, embracing AI not just as a tool but as a fundamental component of their business DNA.

Part Three: Implementing a Robust AI Operating Model

ArgoLong Publishing

Chapter 10

Architecting an AI Operating Model

"An operating model is like the blueprint for a building." —— Wikipedia

H ow can businesses not only navigate but thrive in the rapidly evolving land-scape shaped by artificial intelligence (AI)? What does it take to transform traditional operations into dynamic, AI-driven powerhouses of innovation, efficiency, and customer satisfaction? The journey begins with a vision—a compelling AI vision and a strategy for winning in the digital age. However, envisioning the future is only the first step. The real challenge, and opportunity, lies in making that vision a reality through the meticulous design and implementation of a robust AI operating model.

The transformation from legacy operations to a model where AI is at the core re-quires more than just technological upgrades; it demands a fundamental rethinking of how organizations operate. The stark truth is that many businesses today are ill-equipped for this shift. They are hamstrung by outdated structures, siloed data, and processes that are incompatible with the agility and intelligence AI offers. It's not just about adopting new tools; it's about reimagining the organizational fabric to be AI-first.

This chapter embarks on the critical next step of translating a visionary AI strategy into actionable reality. We dissect the anatomy of an intelligent operating model, essential for any organization aiming to harness AI's transformative potential fully:

- **People**: The cornerstone of change. How do we cultivate an AI-centric culture, secure leadership commitment, attract the right talent, and reor-ganize for agility and innovation?

- **Processes**: The blueprint for action. How do we establish AI governance, incorporate ethical considerations, and set benchmarks that define and measure AI success?

- **Technology**: The engine of transformation. What constitutes a resilient AI tech stack, how do we deploy AI solutions effectively, and what criteria should guide our tool selection?

- **Data**: The fuel for AI innovation. How do we modernize our data infrastructure, craft a data strategy that empowers AI, and ensure our data ecosystem is primed for the next generation of AI applications?

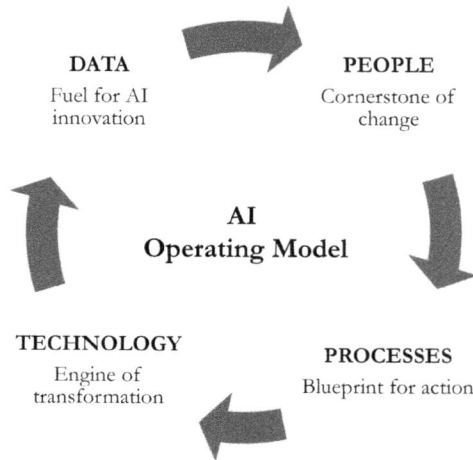

DATA
Fuel for AI
innovation

PEOPLE
Cornerstone of
change

AI
Operating Model

TECHNOLOGY
Engine of
transformation

PROCESSES
Blueprint for action

Figure 10-1: The AI Operating Model with Four Core Components

By addressing these pillars, we offer a comprehensive guide for leaders to transition from traditional to AI-powered operating models. This strategic evolution is pivotal for embedding AI deeply and effectively into the organizational DNA, enabling businesses to achieve their AI ambitions and set new benchmarks in their respective industries.

As we delve deeper, remember that constructing a robust AI operating model is a strategic journey that intertwines technology, people, processes, and data. This chapter aims to illuminate the path forward, providing the insights and framework necessary for leaders to actualize their AI vision and strategy with clarity and purpose.

The AI Operating Model

> **Definition:** The AI Operating Model is defined as an advanced organizational framework that integrates critical components and capabilities essential for the large-scale, responsible, and effective adoption of AI technologies, ensuring alignment with and advancement of an organization's business objectives.

At the heart of an AI-driven transformation lies the AI Operating Model. This model is a holistic aggregation of operational processes, organizational frameworks, technological infrastructures, data strategies, management systems, and evaluative metrics. It is meticulously designed to facilitate the rapid prototyping, validation, deployment, and management of AI technologies within the enterprise environment. By encompassing these elements, the model ensures AI initiatives are seamlessly integrated into the organizational fabric, fostering innovation and driving business growth.

Key Attributes of an Intelligent Operating Model Powered by AI

- **Agile:** The model champions agility through cross-functional teams and decentralized decision-making, coupled with continuous delivery pipelines. This setup empowers organizations to experiment swiftly and adapt quickly to changing market demands and technological advancements.

- **Efficient:** It emphasizes optimized resource allocation and the utilization of shared AI platforms and automated machine learning operations (MLOps). This approach not only ensures economies of scale but also enhances the operational efficiency of AI deployments.

- **Responsible:** Ethical AI is non-negotiable. The model incorporates ethical principles, stringent risk governance, and sophisticated model monitoring mechanisms to ensure the development and deployment of trustworthy AI solutions.

- **Strategic:** A hallmark of the model is its alignment with business priorities, focusing on metrics that accurately track the quality, adoption, and outcomes of AI initiatives. This strategic orientation ensures that AI efforts are directly contributing to the organization's overarching goals.

The AI Operating Model lays down a robust foundation across four critical dimensions: people, process, technology, and data management. This comprehensive base enables organizations to discover, develop, manage, and operationalize AI applications reliably at production scale. It is designed to be dynamic, evolving in response to external regulatory changes, advancements in AI technology, and internal organizational feedback. This continuous evolution accelerates AI adoption and facilitates transparent and accountable practices that generate substantial business and societal value.

By embodying these principles, the AI Operating Model not only equips organizations to navigate the complexities of AI integration but also positions them to leverage AI as a catalyst for sustained innovation and competitive advantage.

Key Components of the AI Operating Model

The AI Operating Model is underpinned by four intricately interwoven capability layers—people, processes, technology, and data. Together, these layers form a comprehensive foundation that is essential for facilitating the adoption of AI across the enterprise, enabling organizations to harness the full potential of artificial intelligence seamlessly and efficiently.

People: The Cornerstone of Change in the AI Operating Model

The people layer focuses on nurturing an AI-ready culture, securing leadership commitment, developing specialized talent, and structural changes essential for an intelligent organization:

The People component of the AI Operating Model is a cornerstone for fostering an ecosystem where artificial intelligence can flourish within an organization. This layer emphasizes the cultivation of an AI-ready culture, the importance of strong leadership commitment, the development of specialized talent, and necessary structural adjustments to support a truly intelligent organization.

AI-Ready Culture: At the heart of fostering an AI-ready culture is the commitment to transparency, accountability, inclusiveness, and purpose. Organizations are encouraged to promote transparency by openly sharing information about the AI systems being developed, thereby increasing awareness and understanding among all stakeholders. Accountability is reinforced through voluntary audits and impact assessments of AI technologies to ensure ethical and responsible use. Inclusiveness is achieved by adopting participatory design thinking approaches in AI development, ensuring that diverse perspectives are considered. Lastly, embedding a sense of purpose into AI solutions by aligning them with societal values and the UN Sustainability Goals ensures that AI initiatives contribute positively to society.

AI Leadership: The role of leadership in an AI-driven organization cannot be overstated. Appointing dedicated AI leaders, such as Chief AI Officers, ensures coordination and oversight of AI initiatives. It is also critical that AI integration into the corporate vision is championed by the CEO and the leadership team, securing a top-down commitment to multi-year investments in AI priorities. This sustained executive commitment is essential for the long-term success of AI strategies.

AI-Ready Talent: Building an AI-ready workforce involves hiring data scientists, machine learning engineers, and MLOps engineers who are pivotal for the development of algorithms and AI systems. Beyond these specialized roles, the creation of ancillary positions such as annotation experts, AI trainers, model interpreters, and AI Ethics officers is crucial for supporting the AI ecosystem. Furthermore, reskilling the broader workforce through immersive AI training programs encourages widespread adoption of AI technologies. Rotational programs across different units facilitate the cross-pollination of AI skills, enhancing the organization's overall AI capabilities.

AI-Ready Structure: Establishing agile, cross-functional enterprise AI Centers of Excellence (CoEs) oriented around business capabilities is a strategic move to centralize AI expertise while aligning with business goals. The launch of AI Guilds focuses on strengthening proficiency in algorithms and data science techniques across the organization. Embedding multidisciplinary AI teams, led by AI Product Managers, within key business units ensures that AI solutions are developed with contextual rigor. Furthermore, reinforcing connections between AI CoEs, Guilds, and business teams through forums and networking events fosters a collaborative AI community.

The People layer of the AI Operating Model is dedicated to building a vibrant and skilled community of AI practitioners. It does so through strategic leadership commitment, comprehensive educational programs, and the establishment of collaborative team structures, all aimed at enabling the broad adoption and ethical application of AI across the organization.

Processes: The Blueprint for Action in the AI Operating Model

The Processes layer of an AI Operating Model is pivotal in ensuring AI adoption is scaled across the enterprise in a manner that is safe, reliable, and responsible. This encompasses the establishment of mature policies, robust control mechanisms, and clearly defined key performance indicators (KPIs), all aimed at fostering a framework where AI can be developed and deployed with integrity.

AI Responsible Framework: Central to this layer is the development of comprehensive guidelines that articulate the organization's commitment to trustworthy, ethical, fair, transparent, and accountable AI. These guidelines are shaped by a blend of external regulations and the core values and brand ethos of the business. They address critical considerations such as privacy, security, safety, transparency, accountability, bias mitigation, interpretability, auditability, and the facilitation of human oversight. The development of these guidelines is a collaborative process, involving consultations with a wide array of stakeholders both within and outside the organization, ensuring a diversity of perspectives is considered.

AI Governance Processes: The governance aspect introduces life cycle policies that cover the entire spectrum of AI system development—from ideation and concept design to data collection, model development, testing, deployment, production monitoring, and eventual retirement. Integral to this process is the conduction of mandatory AI impact assessments. These assessments, which evaluate metrics outlined in the Responsible AI Framework, are a prerequisite before any AI system is approved for use. Additionally, AI systems are subject to voluntary, independent algorithmic audits at regular intervals, alongside periodic model risk reviews utilizing AI explainability and machine learning interpretability techniques. Such measures ensure ongoing oversight and accountability.

AI Success Metrics: Establishing a standardized suite of quantifiable metrics, tailored to the specific business context of the organization, allows for the consistent monitoring of AI model quality, usage levels across business units, and the measurable impact based on the intended purposes and users. These metrics are tracked through a centralized governance platform, offering a comprehensive view of AI adoption rates, system health, and the overall business value being delivered.

At its core, the Processes layer is about instilling a culture of systemic accountability, trust, and transparency within AI initiatives. This is achieved through meticulous policies and control mechanisms that are integrated into the lifecycle of AI development, deployment, and management, ensuring that AI systems are not only effective but also align with the highest standards of ethical responsibility.

Technology: The Engine of Transformation in the AI Operating Model

The Technology layer of the AI Operating Model plays a critical role in enabling organizations to experiment rapidly, develop efficiently, and scale AI applications seamlessly across the enterprise. At the core of this capability is a well-architected AI technology stack, complemented by specialized tools, scalable deployment methods, and a strategic approach to managing the AI lifecycle.

AI Technology Stack: The AI technology stack serves as the foundational framework that supports the entire lifecycle of AI implementation, from data acquisition to the deployment of AI-driven solutions. This stack is meticulously organized into four interconnected layers, each tailored to address distinct aspects of AI technology: the Application Layer, Platform Layer, Model Layer, and Infrastructure Layer. Such a structured approach ensures that organizations have the necessary technological underpinnings to support their AI initiatives effectively.

Generative AI Deployment: The deployment of generative AI, especially utilizing foundation models in vision, speech, and language, is pivotal in propelling AI development forward. These foundational models are instrumental in creating a diverse array of AI applications, enabling rapid progress and customization across sectors. Deployment strategies such as Model Training, Fine-Tuning, Retrieval-Augmented Generation (RAG), and Prompt Engineering are tailored to meet various needs—ranging from developing new applications to enhancing existing models and leveraging real-time information or creatively using established models. Selecting the appropriate deployment strategy is essential for maximizing the AI's impact, customized to specific organizational needs and challenges.

AI Tools Selection Criteria: The selection of AI tools is governed by critical criteria that include the time required to develop the first AI prototype, scalability, collaboration capabilities, and adherence to compliance standards. These criteria ensure that tools not only facilitate rapid development and deployment but also support effective management and monitoring of AI applications, including resource optimization, model drift monitoring, and the incorporation of trust and interpretability by design.

MLOps, AIOps, and AI Lifecycle Management: Understanding and implementing MLOps (Machine Learning Operations), AIOps (Artificial Intelligence Operations), and comprehensive AI Lifecycle Management are essential for the successful deployment and ongoing management of AI systems. These disciplines are constantly evolving, reflecting the rapid pace of innovation in AI technologies and methodologies. They focus on streamlining the AI lifecycle, enhancing operational efficiency, and ensuring AI systems are scalable, robust, and transparent.

As the technological landscape continues to evolve, the emphasis remains on democratizing access to AI technologies through Cloud platforms and optimizing these solutions for efficiency, scalability, robustness, and transparency. The Technology layer, therefore, is not just about selecting the right tools but about architecting a sustainable, agile foundation that enables organizations to harness the full potential of AI.

Data: The Fuel for AI Innovation in the AI Operating Model

The Data component of an AI Operating Model is pivotal in powering AI-driven transformation across an organization. A meticulously crafted data strategy, modern data architectures, and comprehensive data readiness assessments are essential for underpinning enterprise-wide AI initiatives with a solid foundation of high-quality, accessible, and ethically managed data.

Data Strategy for AI Systems: A well-defined data strategy is crucial for the success of AI systems. This includes establishing clear policies for the acquisition, aggregation, and labeling of training data, ensuring compliance with consent and privacy laws, and adhering to ethics guidelines for bias detection and mitigation. Efficient formats for data storage, compression, streaming, and version control are vital to optimize model development processes. Additionally, secure access control policies must be implemented to regulate data access rights across business units, tailored to specific use cases. Responsible data sharing guidelines are also paramount, ensuring that derived data, such as predictions and insights, are shared with transparency and purpose.

Modern Data Architecture: Traditional data architectures, designed for handling static data sets, fall short in meeting the demands of generative AI (Gen AI). Today, dynamic, scalable, and flexible data infrastructures are required to manage the vast volume and variety of data that Gen AI models need for training and deployment. Such a robust and adaptable data architecture is not just a technical necessity but a strategic imperative for leveraging the transformative power of Gen AI. Effective management and processing of multi-structured data not only facilitate the development and refinement of Gen AI models but also their practical application in real-world scenarios.

Data Governance: In the face of exploding data needs and datasets spread across hybrid infrastructures, robust data governance is critical. It ensures that generative AI models have access to accurate and timely inputs, thereby enhancing the quality of outputs. Without proper governance, data strategies may falter, leading to potential chaos and complexity. Comprehensive data governance encompasses Master Data Management, Data Quality Management, Metadata Governance, and adherence to Data Ethics and Regulations, ensuring a structured and controlled data environment.

Data Capabilities: To fully harness the potential of generative AI, organizations must develop specialized data capabilities. This involves focusing on advanced analytics talent, implementing self-service data access platforms, and adopting agile team constructs. Such capabilities enable the organization to manage, analyze, and utilize data more effectively, driving AI innovation.

Ultimately, the Data layer is about efficiently structuring and provisioning data in a way that is both responsible and sustainable. By investing in modern data architectures, organizations can fuel their AI ambitions, fostering an environment ripe for innovation and ensuring their place at the forefront of the AI-driven enterprise landscape.

In conclusion, the harmonious integration of people, processes, technology, and data components forms the backbone of a robust AI Operating Model, enabling organizations to systematically cultivate enterprise intelligence. This cohesive framework not only empowers organizations to drive business value but also ensures that such advancements are achieved responsibly and at scale. By aligning these elements, businesses can navigate the complexities of AI adoption, fostering an environment where innovation thrives, ethical considerations are paramount, and the full potential of AI is unleashed to propel the enterprise forward in a competitive landscape.

Optimizing AI Capability Acquisition for Transformative Success

Definition: An AI capability represents a discrete set of objectives, processes, technologies and talent that collectively enable organizations to responsibly discover, develop, manage and deploy AI systems to achieve business goals.

As organizations navigate the evolving landscape of artificial intelligence, the strategic acquisition of AI capabilities becomes a critical pathway to transformative success. An AI capability is defined as a comprehensive amalgamation of specific objectives, processes, technologies, and talent, orchestrated to empower organizations in the responsible exploration, creation, administration, and implementation of AI systems, thereby facilitating the attainment of business objectives. These capabilities, which encompass specialized roles, expertise, tools, and techniques, are tailored to various phases of the AI model lifecycle, from ideation and data preparation to development, testing, deployment, and monitoring.

For companies seeking to pivot their operational models towards AI, four principal methods of acquiring AI capabilities stand out, each with its unique set of considerations based on speed, control, specificity, competitive advantage, and operational leverage.

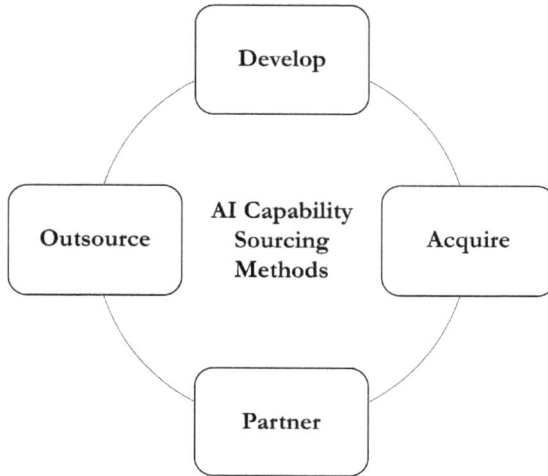

Figure 10-2: The Four Methods of Sourcing AI Capabilities

1. Developing AI In-House: A Strategic Approach to AI Capability Building

Developing AI capabilities in-house involves cultivating a dedicated team of AI experts, resources, and technologies within an organization to create and manage AI systems. This method is particularly advantageous for companies seeking to maintain full control over their AI projects, from conception through deployment, ensuring that the outcomes are highly tailored to their specific operational needs and strategic objectives.

Suitable Situations for In-House AI Development

- **Highly Customized AI Solutions:** When a company requires AI solutions that are closely aligned with its unique business processes, proprietary data, or industry-specific challenges.

- **Competitive Differentiation:** Organizations aiming to develop proprietary AI technologies that serve as a competitive advantage in the market.

- **Sensitive Data Handling:** In scenarios where AI applications need to process highly sensitive or confidential data, keeping development in-house ensures better control over data security and compliance.

- **Long-Term Investment in AI Expertise:** Companies committed to building long-term AI capabilities and embedding AI deeply into their organizational fabric.

When to Use In-House AI Development

- When the organization has access to or is willing to invest in the necessary talent, including AI researchers, data scientists, and engineers.

- If there is a strategic emphasis on building proprietary AI models that require deep integration with existing systems and data.

- When data privacy, security, and regulatory compliance are paramount, necessitating tight control over AI systems.

Advantages of Developing AI In-House

- **Control:** Full authority over the development process, data usage, and customization of AI solutions.

- **Customization:** Ability to tailor AI systems precisely to the organization's specific needs and strategic goals.

- **Competitive Edge:** Potential to develop unique AI capabilities that can serve as a significant differentiator in the market.

- **Data Security:** Enhanced control over data security and compliance with regulations, crucial for sensitive or proprietary data.

Disadvantages of Developing AI In-House

- **Resource Intensity:** Significant investment in talent acquisition, training, and technological infrastructure is required.

- **Time to Market:** Potentially longer development cycles compared to leveraging external solutions or partnerships.

- **Scalability Challenges:** In-house teams may face limitations in scaling AI solutions rapidly without external expertise or partnerships.

- **Risk of Isolation:** Solely focusing on in-house development might limit exposure to external innovations and best practices in the rapidly evolving AI landscape.

Decision Factors

- **Control and Specificity:** The need for high control over AI projects and the specificity of the AI applications to business processes.

- **Speed vs. Competitive Advantage:** Weighing the trade-off between the speed of adoption and the desire for a competitive advantage through unique capabilities.

- **Operational Leverage:** Considering whether the organization has or is willing to develop the operational leverage necessary to sustain long-term AI development in-house.

Developing AI in-house is a strategic choice that offers organizations the benefits of customization, control, and competitive differentiation, albeit with considerations around resource investment, scalability, and the pace of innovation. Organizations must carefully assess their strategic priorities, resource capabilities, and long-term vision for AI adoption when deciding to embark on in-house AI development.

2. Acquiring AI Startups and Specialists: A Rapid Integration Strategy

The strategy of acquiring AI startups and specialists involves identifying and integrating external entities that possess advanced AI technologies, talent, and capabilities into an organization. This approach enables companies to rapidly enhance their AI capabilities, leverage cutting-edge innovations, and fill gaps in their existing expertise or technology stack.

Suitable Situations for Acquisition

- **Immediate Need for Advanced Capabilities:** When an organization needs to quickly integrate advanced AI technologies to stay competitive or enter new markets.

- **Lack of In-House Expertise:** Companies facing challenges in attracting or building the necessary AI talent internally may find acquisitions a faster route to gaining the expertise.

- **Strategic Expansion:** Organizations looking to expand their product offerings or services with AI-driven solutions may find acquisitions a strategic shortcut.

- **Access to Proprietary Technology:** When targeting startups or specialists that have developed proprietary AI technologies which can provide a significant competitive edge.

When to Use Acquisition

- In highly competitive industries where speed to market with AI innovations can determine market leadership.

- When the cost and time investment required to develop similar capabilities in-house are prohibitive.

- To overcome significant barriers to entry in specific AI domains or technologies.

Advantages of Acquiring

- **Rapid Capability Enhancement:** Immediate access to advanced AI technologies and expertise.

- **Speed to Market:** Accelerates the time to deploy AI solutions, offering a faster return on investment.

- **Innovation Access:** Opens doors to innovative AI approaches and technologies not previously available internally.

- **Talent Acquisition:** Brings in specialized talent and leadership with proven AI development and deployment experience.

Disadvantages of Acquiring

- **Cultural Integration:** Challenges in merging the acquired entities' culture with the larger organization, which can impact productivity and morale.

- **Talent Retention:** Difficulty in retaining key personnel from the acquired startups or specialists post-acquisition.

- **Integration Complexity:** The complexity of integrating new technologies with existing systems and processes.

- **Cost:** Acquisitions can require significant upfront investment, with financial risk if the integration or expected synergies fail to materialize.

Decision Factors

- **Urgency and Specificity:** The immediate need for advanced AI capabilities and the specificity of the technology or expertise required.

- **Competitive Advantage:** The potential for the acquisition to provide a significant competitive edge in the marketplace.

- **Operational Leverage:** The readiness of the organization to integrate and leverage the acquired capabilities for operational and strategic advantage.

- **Cultural Fit:** Consideration of the cultural compatibility between the acquiring organization and the target entity to ensure a smooth integration.

Acquiring AI startups and specialists offers a pathway for rapid capability enhancement and innovation access, crucial for maintaining competitive advantage in fast-moving markets. However, it requires careful consideration of integration challenges, cultural fit, and the potential impact on the organization's existing talent and operations. Organizations must weigh these factors against their strategic objectives, market demands, and the urgency of AI capability development to determine if acquisition is the right approach.

3. Partnering with the AI Ecosystem: Leveraging Collaborative Innovation

Partnering with the AI ecosystem involves forming strategic alliances with other companies, research institutions, and technology providers to share resources, expertise, and innovations in AI. This approach enables organizations to tap into a broader pool of AI capabilities and fast-track the development and deployment of AI solutions without the need for extensive in-house development or outright acquisitions.

Suitable Situations for Partnership

- **Access to Complementary Capabilities:** When organizations seek to complement their existing AI capabilities with external expertise or technologies without the need for full integration.

- **Shared Innovation and Risk:** Companies looking to innovate in AI while sharing the development risks and costs with partners.

- **Speed to Market:** Organizations that aim to accelerate their AI initiatives by leveraging the ready-made solutions and expertise available in the ecosystem.

- **Exploratory Initiatives:** For companies experimenting with AI in new areas, partnerships can provide a low-commitment way to explore potential applications and markets.

When to Use Partnerships

- When direct access to cutting-edge AI research, technologies, or platforms is needed to enhance product offerings or operational efficiency.

- To bridge capability gaps in specific AI domains where developing in-house expertise is not feasible or practical in the short term.

- In industries where collaborative models are prevalent, and partnerships can lead to standard-setting or influence market directions.

Advantages of Partnership

- **Flexibility:** Allows for agile engagement with AI technologies and innovations with the ability to adapt partnerships as needs evolve.

- **Reduced Investment:** Lower upfront investment compared to developing in-house capabilities or acquiring companies, with the potential for shared development costs.

- **Broader Innovation Access:** Opens up avenues for collaboration on cutting-edge AI projects and insights from across the ecosystem.

- **Outsourced Workload:** Can offload significant portions of the AI development and operational workload to partners, freeing up internal resources.

Disadvantages of Partnership

- **Vendor Lock-In Risks:** Dependence on external partners for critical AI capabilities could lead to challenges in changing providers or integrating with other systems.

- **Complex Coordination:** Requires managing relationships and coordinating projects across different organizations, which can introduce complexities and slow down decision-making.

- **Potential for Misaligned Objectives:** Partners may have different goals, priorities, or commitments to the partnership, affecting the collaboration's effectiveness.

- **Diluted Control:** Sharing control over AI projects with partners can lead to compromises on project direction, data governance, and intellectual property rights.

Decision Factors

- **Control versus Speed:** Balancing the need for control over AI projects with the urgency to deploy AI solutions quickly.

- **Specificity and Competitive Advantage:** Assessing whether the partnership will provide access to general capabilities or if it can truly offer a unique competitive edge.

- **Operational Leverage:** Evaluating whether the partnership will enhance operational efficiency and scalability without compromising strategic autonomy or incurring excessive coordination costs.

Partnering with the AI ecosystem presents a strategic avenue for organizations to rapidly engage with AI innovations and extend their capabilities through collaboration. It offers flexibility, access to a broad innovation pool, and shared risks, making it an attractive option for companies looking to expand their AI horizons. However, it requires careful management of partnerships, alignment of objectives, and consideration of long-term strategic implications to ensure that such collaborations yield the desired outcomes.

4. Outsourcing AI Capabilities: Flexibility and Scalability at the Forefront

Outsourcing AI capabilities involves engaging external service providers to handle specific AI functions or projects. This strategy allows organizations to leverage specialized expertise and technologies without the need to develop these capabilities in-house or commit significant capital expenditures. It's particularly advantageous for scaling AI initiatives rapidly and accessing high-quality AI solutions with operational flexibility.

Suitable Situations for Outsourcing

- **Resource Constraints:** Ideal for organizations with limited internal resources or expertise in AI and those seeking to implement AI solutions without extensive investment in talent or infrastructure.

- **Scalability and Flexibility:** Suitable for companies requiring the ability to scale AI initiatives up or down based on demand, without the long-term commitments associated with hiring permanent staff or acquiring companies.

- **Non-Core Functions:** Best applied to AI applications that are peripheral to the organization's core business functions, where in-house development may not be justified.

- **Rapid Deployment Needs:** For projects where time-to-market is critical, outsourcing provides access to ready-to-deploy AI capabilities and expertise.

When to Use Outsourcing

- When the organization lacks the in-house expertise to develop or manage AI technologies and hiring or training staff is not viable in the required timeframe.

- For projects or applications where AI is not a core competency of the business, allowing the organization to focus on its primary strengths while still leveraging AI benefits.

- In scenarios requiring quick validation of AI concepts or rapid deployment of AI solutions to respond to market opportunities or competitive pressures.

Advantages of Outsourcing

- **Scalability:** Offers the ability to quickly scale AI efforts up or down based on evolving business needs, without the overhead of maintaining a large in-house team.

- **Cost Efficiency:** Reduces the need for significant upfront investments in technology and talent, shifting AI development costs to operational expenditures.

- **Rapid Deployment:** Enables faster deployment of AI solutions by leveraging external expertise and pre-built technologies.

- **Focus on Core Business:** Allows organizations to concentrate on their core competencies by outsourcing non-essential AI functions.

Disadvantages of Outsourcing

- **Data Privacy and Security Risks:** Outsourcing AI projects involving sensitive data can introduce risks related to data privacy and security.

- **Quality Control:** Maintaining high-quality standards can be challenging when AI development is handled externally, potentially leading to issues with solution effectiveness or integration.

- **Dependency on Vendors:** Reliance on external providers for critical AI capabilities can lead to vendor lock-in, making it difficult to change providers or bring functions in-house in the future.

- **Communication and Coordination Challenges:** Working with external teams may introduce complexities in communication and project coordination, potentially affecting project timelines and outcomes.

Decision Factors

- **Control versus Scalability:** Weighing the need for direct control over AI projects against the benefits of scalability and flexibility offered by outsourcing.

- **Core versus Non-Core Functions:** Determining whether the AI capabilities are central to the organization's business strategy or if they can be effectively managed as external functions.

- **Cost Considerations:** Evaluating the financial implications of developing AI capabilities in-house versus the operational expenditure model of outsourcing.

- **Risk Management:** Assessing the risks associated with data privacy, security, and vendor dependency and implementing strategies to mitigate these risks.

Outsourcing AI capabilities offers a pathway for organizations to rapidly adopt and scale AI technologies with flexibility and cost efficiency. It enables access to specialized expertise and accelerates time-to-market for AI-driven solutions. However, it requires careful selection of partners, clear agreements on deliverables, and robust mechanisms for quality control and data protection to ensure that outsourcing aligns with the organization's strategic objectives and risk tolerance levels.

Criteria	Develop AI In-House	Acquire AI Startups	Partner with AI Ecosystem	Outsource AI Capabilities
Speed	Moderate	High	High	Moderate
Control	High	Moderate	Low	Low
Specificity	High	High	Moderate	Low
Competitive Advantage	Moderate to High	High	Moderate	Low
Operational Leverage	Low	High	Moderate	Low
Pros	Cost-effective, tailored	Quick access, specific solutions	Flexibility, rapid integration	Scalability, low commitments
Cons	Limited depth, external help needed	Integration challenges, talent retention	Vendor lock-in, complex coordination	Data privacy, quality concerns
Best Suited For	Organizations valuing control and specificity	Urgent, specialized capability needs	Rapid integration, less unique advantages	Flexibility, minimal fixed commitments

Table 10-1: The Comparison of Four AI Capability Sourcing Methods

In summary, the selection of an AI capability acquisition strategy demands careful consideration from executives and leaders, who must balance the benefits and challenges across different approaches with the broader objectives and characteristics of their organization. Factors such as speed, control, specificity, competitive advantage, and operational leverage are pivotal in guiding this decision-making process. The most suitable AI sourcing strategy will be one that aligns with the organization's strategic vision, its pace of innovation, the availability of talent, its tolerance for risk, and its cultural dynamics. This careful alignment ensures that the chosen method not only meets immediate needs but also positions the organization for long-term success and competitiveness in the rapidly evolving technological landscape.

Strategic Allocation of AI Capabilities in Business Operations

The integration and management of AI capabilities within an organization are critical processes that must be thoughtfully aligned with the specific types of value these capabilities provide and their relevance to different business units. This strategic alignment involves discerning the most effective segments of the organization to handle various AI functions, categorized by their demand-side or supply-side advantages and their proximity to core business activities.

Shared AI Capabilities focus on supply-side benefits, such as operational efficiency and resource optimization. These capabilities are typically centralized to maximize scale and cost-efficiency, potentially leveraging offshoring or outsourcing. For example, a global retail chain might use centralized AI platforms for inventory management, employing MLOps to streamline supply chain logistics and model governance to ensure compliance across markets. This approach is ideal for standardized, repetitive tasks across the enterprise, such as automated customer inquiries handling through chatbots.

Strategic AI Capabilities are demand-side oriented, crucial for guiding organizational strategy, branding, and high-value initiatives. Located near the heart of the organization, these capabilities enable swift, strategic decision-making. An example is a tech company using AI-driven analytics for corporate strategy, where AI tools analyze market trends and consumer behavior to inform product development and marketing strategies. High-level AI project steering in such companies often involves direct oversight from top executives, ensuring AI initiatives align with the company's strategic goals.

Business-Specific AI Capabilities are intimately linked with particular lines of business, focusing on demand-side aspects and expert abilities such as R&D and product marketing. These capabilities are embedded within individual business units to ensure tailored solutions. For instance, a pharmaceutical company might integrate AI in its R&D unit to accelerate drug discovery, while its marketing department uses AI to tailor advertising campaigns to specific demographics. Such embedded AI capabilities ensure direct impact on business outcomes, leveraging deep domain expertise.

Specialist AI Capabilities stand out for their high specialization, encompassing both demand and supply-side aspects, such as innovation and technical sales. Organized often as centers of excellence, these capabilities combine deep technical expertise with innovation. A notable example is an automotive company establishing a center of excellence for AI-driven autonomous driving technologies, focusing on advanced AI research and algorithmic innovation to stay at the forefront of the autonomous vehicle market. These centers are pivotal for areas requiring continuous learning and collaboration with the broader AI and tech ecosystem.

AI Capability Type	Nature	Allocation	Examples	Strategy
Shared AI Capabilities	Supply-side, operational efficiency	Centralized, can be offshored or outsourced	AI platforms, MLOps, model governance	Standardized tasks, enterprise-wide efficiency
Strategic AI Capabilities	Demand-side, strategic value	Close to headquarters and executives	AI-driven corporate strategy, high-level AI projects	Talent-driven, high-value initiatives
Business-Specific AI Capabilities	Demand-side, expert abilities in business lines	Embedded in business units	Marketing AI, sales AI, product development AI	Tailored to specific business units and their goals
Specialist AI Capabilities	Both demand and supply-side, innovation	Centers of excellence	Advanced AI research, algorithmic innovation	Continuous learning, partnerships, innovation-focused

Table 10-2: The Comparison of Four AI Capability Types

Furthermore, the delivery approach for these AI capabilities should intrinsically align with the company's overarching strategy. For example, a company with straightforward customer support requirements might opt for centralizing and automating most functions, providing specialized support only when necessary. In contrast, a business dealing with complex products could regard customer support as a specialist capability, necessitating a team of highly skilled staff adept at solving diverse problems and identifying upselling opportunities.

By strategically aligning the allocation of AI capabilities with business imperatives, companies can refine their operating models to enhance efficiency, drive innovation, and improve market adaptability. This strategic alignment ensures that AI initiatives are not only technologically advanced but also deeply integrated with and responsive to the specific needs and goals of the business.

Embracing AI-Native Thinking in Operating Models

AI-native operating models herald a significant transformation in the way businesses structure and manage their operations, signaling a shift from mere incorporation of AI technologies to a fundamental reevaluation of business processes, decision-making frameworks, and strategic orientations through the lens of AI's expansive capabilities and potential.

Agility and Flexibility at the Core: Central to AI-native operating models is the prioritization of agility and flexibility, enabling organizations to swiftly adapt to technological advancements and market dynamics. Emphasizing scalable AI solutions and the ongoing refinement of AI-driven processes ensures that businesses are well-equipped to seize new opportunities and tackle emerging challenges with promptness and efficiency.

Revolutionizing Core Business Processes: AI's impact on core business processes is profound, automating routine tasks, streamlining workflows, and enhancing decision-making capabilities. This revolution leads to heightened operational efficiency, significant cost reductions, and increased accuracy in essential operations, marking a departure from traditional operational paradigms.

Leveraging Real-Time Data Analytics for Decision Making: The advent of AI in performing real-time data analytics transforms decision-making processes. Instant insights allow for agile responses to market fluctuations, providing businesses with a considerable competitive advantage. Such capabilities necessitate a restructured operating model that can fully exploit real-time analytics for strategic benefit.

Personalizing Customer Experience: Through AI's advanced data analysis, businesses gain deep insights into customer preferences, facilitating highly personalized customer experiences. The integration of AI into customer interactions ensures services and communications are tailored and responsive, setting new standards in customer engagement.

Supply Chain and Logistics Optimization: AI redefines supply chain management by improving demand forecasting, inventory management, and logistical efficiency. Such enhancements not only elevate operational effectiveness but also heighten customer satisfaction through better product availability and expedited deliveries.

Advancing Predictive Maintenance and Quality Control: In sectors like manufacturing and services, AI plays a pivotal role in enabling predictive maintenance and ensuring consistent quality control. This proactive stance minimizes downtime and guarantees product excellence, which are vital for sustaining brand reputation and customer loyalty.

Enhancing Collaboration and Communication: AI technologies facilitate improved collaboration and communication across organizational boundaries, dismantling silos and fostering cohesive teamwork. This is particularly invaluable in complex projects that require a diverse range of expertise.

Ensuring Ethical AI Utilization: The integration of AI into business operations mandates a steadfast commitment to ethical guidelines and compliance standards, covering data privacy, bias mitigation, and regulatory adherence. This ethical foundation is crucial for building trust and credibility in AI deployments.

Transforming the Workforce: As AI automates standard tasks, the emphasis shifts towards roles centered around strategic insight, creativity, and complex problem-solving. This evolution calls for a comprehensive redefinition of job roles, performance metrics, and training paradigms, aligning the workforce with the demands of an AI-driven business environment.

Commitment to Continuous Improvement: An AI-native model thrives on perpetual improvement and learning, requiring organizations to stay current with AI advancements and seamlessly integrate emerging technologies and methodologies. This commitment is essential for sustaining a competitive edge in an increasingly AI-dominated business landscape.

Adopting an AI-native operating model is a holistic endeavor that extends beyond technological adoption. It demands a reimagined approach to business operations, where agility, ethical deployment, and a culture of continuous learning are foundational elements. By fully embracing this dynamic and forward-thinking model, organizations position themselves to harness the transformative power of AI, driving significant change and achieving sustainable growth in the digital era.

Crafting an AI-Enhanced Operating Model for Tomorrow's Enterprise

Embarking on the journey to establish an AI-driven operating model across an enterprise is both a challenging and essential pursuit. It requires the careful alignment of people, processes, technology, and data platforms to leverage the full potential of AI, mitigating risks and maximizing benefits in the process. This chapter has laid out a sophisticated framework and strategic playbook for leaders dedicated to evolving their organizations for the AI age, focusing on several critical areas:

Structured Guidance: We've provided a thorough blueprint of the indispensable components required for building an intelligent enterprise, including human capital, policy frameworks, technological infrastructure, and data architectures. These elements are foundational for any organization aspiring to harness AI's capabilities effectively.

Strategic Approaches: Through a nuanced examination of different strategies for acquiring AI capabilities—weighing the pros and cons of developing in-house, purchasing, or partnering—we've outlined options tailored to meet organizational objectives and innovation timelines. This analysis aids in making informed decisions that align with long-term visions and immediate needs.

Actionable Steps for Leadership: Guidance has been offered on how executives can steer their organizations from conventional operational frameworks to those enhanced by AI. This transition involves strategic initiatives in talent acquisition and development, ethical governance, responsible data management, selective tool adoption, and forging meaningful partnerships.

Continuous Evolution: The chapter emphasizes that adopting an AI-enabled operating model is an iterative process, necessitating regular revisions in light of changing regulations, technological breakthroughs, and feedback from within the organization. This dynamic approach ensures that the operating model remains relevant and effective.

Synergistic Integration: We advocate for a holistic integration of all components of the operating model, encouraging organizations to adopt a comprehensive and coordinated approach to AI adoption. Such integration is crucial for accelerating the successful implementation of AI technologies and practices.

In essence, the creation of an AI-driven operating model is an ongoing endeavor that demands constant attention, innovation, and adaptation. By adopting a holistic strategy that marries AI technology with deep human insights, organizations can secure their place as leaders in efficiency and innovation in the digital era.

The forthcoming chapters will explore in greater detail the four foundational elements crucial to an AI-driven operating model. These sections aim to furnish leaders and organizations with in-depth knowledge and practical advice, enabling them to effectively tackle the intricacies of embedding AI into their operational frameworks, paving the way for a future-proof enterprise.

Chapter 11

AI-Powered Workplace: A Blueprint for the Next-Gen Workforce and Culture

"A strong economy begins with a strong, well-educated workforce." — Bill Owens

I n the dawn of a new era, where the boundaries between human intellect and artificial intelligence blur, how does the future of work reshape itself? Imagine a workplace where AI doesn't stand as a rival but as a partner, amplifying human potential. This is not a distant reality or a scene from a science fiction novel; it's the transformation unfolding right before our eyes, as delineated by recent MIT Sloan studies. The narrative that once painted AI as a formidable job-stealer, preying on tasks deemed mundane, has evolved into a more nuanced and optimistic dialogue. It's a narrative of augmentation, not replacement, of collaboration, not competition.

As we stand at this pivotal intersection of labor and technology, it's crucial to peel away the layers of fear and apprehension to reveal the true essence of AI's role in the workplace. This chapter aims to demystify the complexities of AI integration and chart a path forward for organizations and employees alike.

AI's Multi-Dimensional Impact

AI is redefining productivity and efficiency, not by rendering human efforts obsolete, but by enhancing them. Its unparalleled ability to process vast datasets and execute repetitive tasks with precision liberates the workforce to pursue creative and strategic endeavors. Moreover, AI's knack for predictive analysis and decision-making illuminates paths that were once obscured by the complexity of data. This capability is a boon for sectors like finance, marketing, and strategic planning, where foresight can translate into competitive advantage.

The personalization potential of AI marks a revolution in how products and services are tailored, offering customization at a granular level that was previously unfathomable. Beyond optimizing the existing, AI paves the way for innovation, birthing new industries and job roles, thereby enriching the job market.

Navigating the Challenges

Yet, the path to AI integration is strewn with challenges. The shift necessitates a recalibration of skills, a cultural metamorphosis within organizations, and a reimagining of traditional job roles. The advent of AI demands a workforce that is not only technically adept but also versatile, innovative, and continuously learning.

This chapter is not just an exploration of AI's potential to transform workplaces but also a guide to embracing this change. It underscores the importance of preparing for a future where human ingenuity and artificial intelligence coalesce, fostering an environment of productivity, innovation, and mutual growth. As we venture into this new age, the question remains: How can we ensure that the AI-powered workplace becomes a realm of opportunity, collaboration, and enhanced job satisfaction for all?

The journey into the AI-powered workplace is both an opportunity and a challenge. It calls for a paradigm shift in how we perceive work, value human contributions, and harness the power of artificial intelligence. By navigating this transition thoughtfully, we can unlock a future where the workplace is not just more efficient and innovative but also more human-centric and fulfilling.

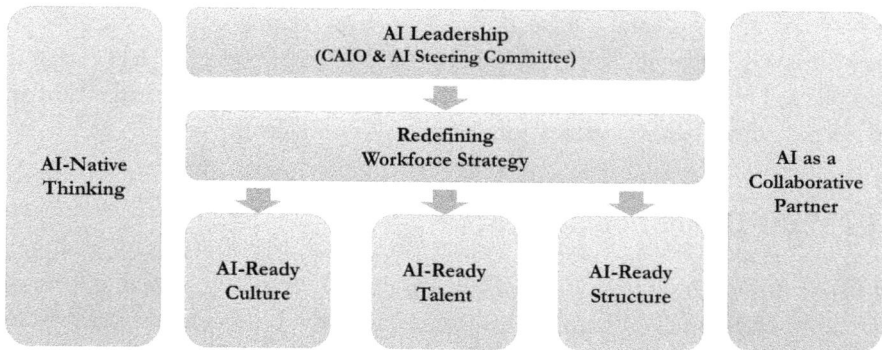

Figure 11-1: The Blueprint for the AI-Powered Workplace

AI as a Collaborative Force in the Workplace

In today's rapidly evolving work environment, the emergence of artificial intelligence (AI) as a collaborative force is reshaping the traditional dynamics of teamwork. Instead of perceiving AI as a competitor to human roles, envisioning it as a partner that amplifies and extends human capabilities offers a more progressive and fruitful perspective. This new paradigm thrives on the symbiotic strengths of human ingenuity and AI's computational power, forging a partnership that enhances the collective output of teams.

Synergizing Human Creativity and AI's Analytical Prowess

Cognitive Synergy: AI's proficiency in digesting vast amounts of data complements the human capacity for creativity, critical thinking, and contextual understanding. For example, in creative sectors, AI can generate options informed by data trends, while humans infuse these suggestions with depth and relevance, resulting in impactful and resonant work.

Collaborative Problem-Solving: AI's ability to pinpoint challenges and areas for improvement, potentially unnoticed by humans, combined with human insight, leads to more nuanced and effective solutions to intricate issues.

Revolutionizing Communication: AI is transforming how we communicate, with tools ranging from intelligent assistants that streamline scheduling to translation technologies that bridge linguistic divides, thereby enhancing workplace efficiency and cohesion.

AI's Role in Support and Enhancement

Streamlining Administrative Functions: By automating routine tasks such as scheduling and data management, AI liberates human employees to pursue more meaningful and strategic initiatives.

Personalized Learning and Development: AI-driven learning platforms adapt to the unique preferences and speeds of learners, making education more engaging and efficient, crucial for continuous professional growth.

Promoting Health and Well-being: AI's capability to monitor and analyze workplace well-being metrics offers invaluable insights into stress, workload equilibrium, and health, facilitating a more balanced and productive work environment.

Navigating the Human-AI Collaboration Landscape

Embracing the collaborative potential of humans and AI also involves addressing certain challenges:

Trust and Reliability: Establishing trust in AI systems is essential for their acceptance. Workers must be confident in the AI's reliability, fairness, and transparency.

Ethical Deployment: The expanding role of AI in decision-making and privacy necessitates a strong ethical framework to ensure its responsible use.

Workflow Integration: The introduction of AI into established workflows demands careful planning, training, and support to ensure a seamless transition and to maximize the benefits of AI-human collaboration.

The vision of AI as a collaborative partner in the workplace heralds a new era of enhanced efficiency, creativity, and job satisfaction. By recognizing and nurturing the complementary strengths of AI and human employees, organizations can unlock unprecedented levels of innovation and productivity. This collaborative future not only optimizes work processes but also enriches the work experience, paving the way for a more dynamic and fulfilling professional landscape.

The Rise of AI Leadership

Artificial Intelligence (AI), recognized as a versatile and foundational technology, has emerged as a pivotal force of transformation across various sectors. To harness this power effectively and ethically, the introduction of specialized AI leadership roles, such as the Chief AI Officer (CAIO) and AI Steering Committees, has become a strategic imperative for forward-thinking organizations.

The Chief AI Officer: Spearheading AI Strategy

The CAIO, a pivotal figure within the C-suite, orchestrates the organization's AI initiatives from conception to implementation. This role transcends technical management, embodying a strategic vision that integrates AI with the company's overarching objectives. Responsibilities of a CAIO include:

- **Strategic Vision and Implementation:** Crafting and executing a comprehensive AI strategy that enhances business operations and drives innovation.

- **Leadership in Team Dynamics:** Guiding multidisciplinary teams of AI experts, data scientists, and developers to foster groundbreaking AI solutions in harmony with IT and other business units.

- **Ethics and Governance:** Championing ethical AI practices, ensuring compliance with relevant laws and standards, and developing frameworks for accountability and transparency in AI applications.

- **Stakeholder Engagement:** Acting as an AI ambassador, the CAIO clarifies AI's role and potential benefits, addressing any concerns and fostering an AI-literate culture across the organization.

AI Steering Committees: Facilitating Cross-Functional Collaboration

To complement the CAIO, many organizations have established AI Steering Committees. These cross-functional groups ensure that AI strategies are congruent with corporate values and goals, embracing a holistic approach to AI governance. Key functions include:

- **Collaborative Leadership:** Harmonizing efforts across departments, the committee navigates the multifaceted perspectives and objectives inherent in diverse teams.

- **Ethical Oversight:** It safeguards ethical AI practices, monitors compliance with regulatory standards, and proactively addresses challenges related to data privacy and algorithmic bias.

- **Change Management:** Steering the organization through the AI transition, the committee oversees employee engagement, training, and the cultural shift towards AI integration.

- **Strategic Decision-Making:** The committee aids in crucial decisions regarding AI investments, prioritizing projects, and allocating resources to align AI endeavors with strategic business objectives.

The Imperative for AI Leadership in the Modern Enterprise

The establishment of a CAIO and an AI Steering Committee signifies an organization's dedication to leveraging AI's capabilities responsibly. These leadership roles provide focused strategic direction and ensure organization-wide coherence in AI deployment.

Considerations for Instituting AI Leadership

- **Organizational AI Maturity:** The decision to introduce AI-specific leadership roles is contingent on the organization's progress and ambitions within the AI domain.

- **Industry Demands:** Sectors with high reliance on data analytics, such as healthcare and finance, may find immediate value in dedicated AI governance structures.

- **Resource Commitment:** The viability of these roles also depends on the organization's resource availability and its commitment to investing in AI technology.

As AI continues to redefine the business environment, the necessity for specialized AI leadership roles like CAIOs and AI Steering Committees becomes increasingly vital. These roles not only champion the strategic and ethical deployment of AI but also ensure its integrated application across business functions. Moving forward, as AI technologies advance, the demand for such dedicated leadership is set to grow, guiding enterprises through the dynamic landscape of AI-enhanced business operations.

Redefining Workforce Strategy for the Generative AI Revolution

In the dawn of the generative AI revolution, businesses are at a pivotal juncture where reimagining talent strategy is not just beneficial but essential. This technological advancement compels a shift in leadership mindset: from perceiving roles as potentially automated and replaceable to valuing unique human contributions within an evolving business landscape. A forward-looking, adaptable talent strategy that harnesses the power of AI collaboration and prepares the workforce for an AI-enhanced future is imperative.

Customized Impact Analysis for Strategic Adaptation: Generative AI's impact on organizations will vary, necessitating a tailored strategy that considers each enterprise's unique context. This strategy encompasses assessing demand drivers for products and services alongside technological factors influencing AI adoption and integration. Understanding these drivers is critical for determining how to scale operations and where to channel investments in talent and technology, thereby influencing work organization, role design, and workflow innovation.

Redefining Job Roles Amid AI Integration: Generative AI is poised to redefine a broad spectrum of tasks, from creative content generation to coding, impacting job roles across organizations in unpredictable ways. The challenge for leaders is to anticipate how this technology will be deployed—to enhance productivity, eliminate specific roles, or pioneer new services. Such foresight is vital for strategic planning, ensuring teams are prepared for evolving workloads and the organization remains poised to meet changing market demands.

Strategic Demand Forecasting and AI Application: Proactive demand forecasting is essential for navigating the generative AI landscape. This foresight into product or service demand will directly inform operational scaling and investment strategies. Following this, envisioning the specific applications of generative AI technology becomes crucial. Leaders must assess whether AI will supplant certain jobs, necessitate new competencies, or become a linchpin for competitive advantage.

Evolving Roles and the Emergence of New Specializations: As generative AI integrates into business operations, the transformation of roles is inevitable, possibly leading to a streamlined workforce where fewer individuals are required for the same output. This evolution does not inherently signal widespread job losses but indicates a shift towards redesigned, multifaceted roles and the gradual emergence of new, highly specialized positions. These future roles will blend deep business acumen with technological fluency, epitomized by roles such as executive-level business architects or interdisciplinary teams blending diverse skills.

The era of generative AI demands a robust, visionary talent strategy that anticipates the dual need to transform existing roles and foster new specializations. By proactively adapting to this dynamic landscape, organizations can ensure their competitiveness and agility in a world increasingly shaped by AI innovations. This strategic pivot not only prepares businesses for the immediate impacts of AI but also positions them to lead in the creation of future industries and job categories.

AI-Ready Culture: A Blueprint for Successful Integration

The transition into the era of Artificial Intelligence (AI) demands more than just technological adoption; it requires the cultivation of an AI-Ready Culture. Such a culture is characterized by a workforce that not only understands and accepts AI but is also actively engaged in its deployment. This environment is pivotal for sparking innovation, enabling data-driven decision-making, and fostering a commitment to continuous learning—elements crucial for maximizing the benefits of AI.

Foundational Pillars of an AI-Ready Culture

- **Trust and Transparency:** Establishing a foundation of trust is essential, ensuring employees believe in the organization's capability to develop competent AI systems that enhance rather than undermine their roles. Transparency about AI's development process and its applications reinforces this trust.

- **Accountability and Ethical AI Use:** Embedding accountability through regular audits and impact assessments guarantees the responsible utilization of AI. Open dialogue about ethical considerations and the consequences of AI applications underscores this commitment.

- **Inclusivity in AI Development:** Embracing inclusivity through participatory design approaches ensures that AI solutions benefit from a wealth of perspectives, enhancing their relevance and effectiveness.

- **Alignment with Societal Values:** Purpose-driven AI solutions, aligned with societal values and global sustainability goals, ensure that the technology's impact transcends organizational boundaries, contributing to broader societal well-being.

Strategies for Fostering an AI-Ready Culture

- **Leadership Advocacy:** Visible support and advocacy from leadership signal a strong commitment to AI, setting a precedent for organization-wide engagement.

- **Education and Training:** Providing comprehensive training programs demystifies AI for employees, clarifying its impact on their roles and the broader organizational context. This should include insights into AI's capabilities, limitations, and ethical use.

- **Encouraging Continuous Learning:** Promoting a culture of continuous education on AI trends and developments spurs innovation and creative problem-solving.

- **Fostering Collaboration:** A collaborative culture that encourages cross-departmental interaction and open inquiry about AI fosters broader understanding and acceptance.

- **Highlighting AI Successes:** Sharing stories of successful AI projects within the organization can illustrate its tangible benefits, encouraging wider acceptance and enthusiasm.

- **Adapting Roles for AI:** As AI assumes routine tasks, redefining roles to emphasize strategic, creative, and complex tasks can help employees see how their contributions evolve alongside AI.

- **Promoting Experimentation:** An environment that supports AI experimentation, backed by necessary resources and a flexible approach, nurtures a dynamic and adaptable workforce.

- **Integrating AI Ethics:** Making AI ethics a cornerstone of the organizational culture ensures a workforce committed to responsible AI use.

Evaluating the success of an AI-ready culture involves metrics such as employee engagement levels in AI training, the implementation rate of AI initiatives, and feedback on AI's organizational impact. Overcoming obstacles like change resistance and job security concerns necessitates clear, empathetic communication about how AI augments, rather than replaces, human capabilities.

In summary, building an AI-ready culture is not just a strategic advantage but a necessity for organizations aiming to thrive in the AI era. Through education, leadership engagement, inclusivity, and a steadfast commitment to ethical AI, organizations can lay the groundwork for a future where AI acts as a catalyst for innovation and operational excellence. As AI technologies continue to advance, nurturing an AI-ready culture will be critical for maintaining competitiveness and driving innovation in an ever-changing business landscape.

AI-Ready Talent: Cultivating a Workforce for AI Innovation

Crafting a workforce that is not only ready but also enthusiastic about embracing Artificial Intelligence (AI) involves more than just imparting technical knowledge. It requires a comprehensive strategy for education and professional development that aligns human intelligence with the capabilities of AI technologies, fostering an environment ripe for innovation across various sectors.

Expanding AI Literacy Across the Board: Embarking on a journey toward AI fluency entails the development and implementation of extensive AI literacy programs. These initiatives should cover the intricate technicalities of AI, as well as its ethical, social, and economic ramifications. Ensuring widespread accessibility to AI and machine learning education across every stage of learning—from primary schools to continuous professional development programs—is fundamental to building a foundationally strong AI-savvy workforce.

Elevating AI-Complementary Human Skills: As AI systems increasingly automate routine tasks, the premium on human-centric skills such as critical thinking, creativity, emotional intelligence, and problem-solving is set to rise. Educational structures must prioritize these skills, enabling individuals to work in tandem with AI technologies effectively.

Encouraging a Culture of Lifelong Learning: Adopting a lifelong learning ethos within the workplace is crucial for keeping pace with the swift evolution of AI technology. Organizations should provide ample resources and opportunities for employees to continually refine and augment their skills in alignment with the latest AI developments.

Demonstrating AI's Cross-Disciplinary Potential: To fully harness AI's transformative impact, it's essential to illustrate its applicability across a wide array of fields, including but not limited to healthcare, finance, and manufacturing. Providing a deep contextual understanding of how AI can address specific industry challenges enriches its application and maximizes its benefits.

Instilling Ethical Principles in AI Deployment: Integrating ethical considerations into AI training—addressing data privacy, algorithmic bias, and the broader impact on society—is critical. Building a workforce that is both technically proficient and ethically grounded ensures the responsible deployment of AI technologies.

Designing Collaborative Human-AI Workspaces: Creating work environments that support and enhance human-AI collaboration, coupled with encouraging team-led exploration of AI's potential, is key to leveraging the collective strengths of both human and artificial intelligences.

Overcoming Workforce Development Hurdles: Addressing the challenges of AI workforce development involves targeted reskilling and upskilling efforts, designed to bridge the skills gap and anticipate future employment trends. Ensuring equitable access to AI training and opportunities, especially for underrepresented groups, is vital for nurturing a diverse and inclusive AI future.

Fostering a Diverse AI Talent Ecosystem: The evolving AI landscape demands a broad spectrum of specialized professionals—from data scientists to ethical AI overseers. Organizations must invest in training that spans the gamut of AI knowledge, from foundational principles to cutting-edge applications, to cultivate a workforce capable of driving AI innovation.

Promoting Interdisciplinary Knowledge Sharing: Implementing cross-functional rotation programs can facilitate the widespread adoption of AI knowledge and skills, promoting a culture of interdisciplinary collaboration and continuous learning.

Balancing AI Talent Management Approaches: Deciding between a centralized approach to AI talent management and a decentralized model involves weighing the benefits of uniform strategy and training standards against the need for innovation and customized solutions.

Attracting and Retaining Premier AI Talent: To attract the best AI professionals, organizations must clearly communicate their data strategies and advancement opportunities. Expanding the search to include untapped talent pools and tailoring recruitment processes can significantly enhance hiring success rates.

Preparing for the AI revolution entails more than just technological readiness; it requires a strategic, holistic approach to workforce development that emphasizes continuous learning, ethical engagement, and innovative thinking. As AI continues to reshape the landscape of industries, the readiness and adaptability of the workforce will be paramount in harnessing the full potential of AI technologies for future growth and innovation.

AI-Ready Organizational Structure

In the fast-paced realm of Artificial Intelligence (AI), crafting an organizational structure that not only accommodates but thrives on AI innovation is crucial. This involves setting up agile, multifaceted teams and centers dedicated to AI excellence, ensuring that AI initiatives are both strategic and seamlessly integrated across business operations.

Enterprise AI Centers of Excellence: Catalysts for Innovation

AI Centers of Excellence (CoEs) act as the driving force in organizations for AI initiatives. These centers, comprised of cross-functional teams, focus on specific business capabilities and foster a culture of innovation and experimentation. They provide the necessary guidance, best practices, and expertise to ensure AI projects align with business goals and are executed effectively. CoEs can be pivotal in scaling AI solutions across the enterprise, ensuring consistency and efficiency in AI deployments.

AI Tiger Teams and Agile Scrum Teams for AI Pilot Work

AI tiger teams or agile Scrum teams are essential for pilot AI projects. These teams are typically small, cross-disciplinary groups tasked with rapidly developing and testing AI solutions in a controlled environment before wider roll-out. Their agility allows for quick iteration and adaptation based on feedback, making them ideal for exploring new AI applications and innovations.

Integrating AI Expertise Directly into Business Units

Placing multidisciplinary AI teams within key business units, led by AI Product Managers, ensures that AI solutions are developed with contextual rigor. This embedded structure allows for a deeper understanding of specific business challenges and opportunities, leading to more tailored and effective AI solutions. AI Product Managers play a crucial role in bridging the gap between technical AI capabilities and business needs.

AI Guilds: Forging Technical Mastery

AI Guilds are communities within organizations focused on strengthening proficiency in algorithms, data science techniques, and other AI-related skills. These guilds provide a platform for continuous learning, knowledge sharing, and collaboration among AI practitioners. They play a crucial role in keeping the workforce up-to-date with the latest developments in AI and data science, thereby enhancing the organization's overall AI capabilities.

Reinforcing Connections through Forums and Networking Events

To maximize the impact of AI CoEs and Guilds, organizations should facilitate regular forums and networking events. These events serve as platforms for knowledge exchange, problem-solving, and collaboration between different AI teams and business units. They help in aligning AI initiatives with business strategies and fostering a shared understanding of AI's role and potential across the organization.

Choosing Between Centralized and Decentralized AI Organizational Models

- **Centralized Model:** This model centralizes AI governance, often within a dedicated department or CoE, ensuring uniform AI strategy and policy execution across the enterprise. It's particularly effective in safeguarding AI expertise quality and aligning AI efforts with overarching organizational directives.

- **Decentralized Model:** Conversely, a decentralized framework distributes AI resources and initiatives across various business units, enhancing flexibility and allowing for bespoke solutions to department-specific challenges. This model is conducive to fostering innovation and agility at the local level.

The choice between centralized and decentralized models depends on the organization's size, culture, and business goals. While a centralized model offers more control and consistency, a decentralized approach can drive innovation and responsiveness. Some organizations opt for a hybrid model, combining elements of both to balance control with flexibility.

In summary, building an AI-ready organizational structure involves creating agile, cross-functional teams, embedding AI expertise within business units, fostering continuous learning through guilds, and maintaining robust communication channels. The choice between centralized and decentralized structures should align with the organization's overall strategy and goals. As AI continues to evolve, organizations must adapt their structures to leverage AI effectively, ensuring they remain competitive and innovative in the AI-driven business landscape.

Maximizing Productivity in High-Skill Domains through Generative AI

Generative AI, particularly through the advancements in Generative Pretrained Transformers (GPT), is revolutionizing the efficiency and output of highly skilled professionals. Research spotlighted by MIT Sloan suggests a striking productivity increase of up to 40% in tasks augmented by generative AI, compared to those conducted without it. This leap in performance is primarily due to AI's proficiency in producing outputs that closely mimic human creativity, thereby significantly aiding in tasks that require quick ideation and the crafting of compelling narratives.

Harmonizing AI Potentials with Human Expertise

The key to harnessing generative AI effectively lies in recognizing its boundaries. Instances where reliance on AI overshadows critical human judgment can lead to a dip in performance, marking the importance of discerning the "jagged technological frontier" of AI. This underscores the necessity for professionals to apply their expertise in tandem with AI, critically assessing AI-generated outputs to ensure they meet the specific demands of their tasks.

Strategies for Integrating Generative AI into Professional Workflows

- **Interface Design and Effective Onboarding:** The architecture of AI interfaces is critical in shaping how intuitively professionals can leverage AI tools. Interfaces that clearly delineate AI's strengths and limitations, coupled with comprehensive onboarding processes, can significantly enhance user experience and productivity.

- **Reconfiguring Roles and Aligning Tasks:** Identifying tasks that generative AI can optimally perform is essential for its effective integration into professional settings. Organizations should foster a culture of experimentation to discover the most efficacious ways to incorporate AI into various roles and responsibilities.

- **Cultivating Accountability in AI Usage:** Establishing a culture where professionals are accountable for how they utilize AI ensures its role as an augmentative tool rather than a crutch. This involves understanding AI's applications and being able to articulate the rationale behind AI-assisted decisions.

Generative AI as a Catalyst for Upskilling and Collaborative Learning

An intriguing finding from the study is the more pronounced benefit of AI on enhancing the performance of workers with lower initial skill levels, suggesting generative AI's potential as a powerful upskilling tool. Implementing peer training programs where more experienced professionals guide their colleagues in adapting to AI can amplify this upskilling effect. Recognizing and incentivizing these peer mentors are critical steps in fostering a supportive and collaborative learning environment.

Generative AI emerges as a transformative force for high-skill professions, offering substantial productivity gains and creative capabilities. Its successful deployment, however, hinges on a thoughtful integration strategy that respects AI's capabilities while emphasizing human expertise and judgment. As the landscape of AI continues to evolve, striking the right balance between leveraging AI's computational strengths and nurturing human insight will be paramount in unlocking the full potential of generative AI in enhancing workplace productivity.

Leveraging Generative AI to Equalize and Enhance Workforce Skills

Generative AI, particularly through advanced large language models, is revolutionizing the workplace by significantly enhancing the productivity of less experienced workers. A pivotal study from MIT Sloan underscores how generative AI technologies are not just tools for efficiency but also powerful allies in rapidly upskilling employees, marking a notable shift in the landscape of workforce development.

The Equalizing Influence of Generative AI on Productivity

Empowering New Entrants: The study reveals a compelling narrative where contact center agents equipped with conversational AI assistants saw a productivity leap of 14%. This boost was especially significant among the novices or those with limited skills, underscoring generative AI's capacity to fast-track proficiency, a process traditionally measured in months or years.

Bridging the Skill Gap: Generative AI emerges as a great leveler in the workforce, diminishing the divide in productivity that typically separates seasoned workers from their less experienced counterparts. By providing tailored support, generative AI narrows the efficiency gap, making expertise more universally accessible and reducing skill-based inequality in the workplace.

Navigating the Implications and Ethical Considerations

Balancing Productivity with Personal Development: While the immediate productivity gains from generative AI are clear, it's crucial to balance these advances with authentic learning and skill acquisition. Dependence on AI for decision-making without grasping the rationale could hinder deeper learning and long-term professional growth.

Who Gains from AI-Driven Productivity?: The distribution of benefits from productivity enhancements raises important questions. Developing equitable frameworks to ensure that the gains are not disproportionately skewed but rather beneficial across the spectrum of stakeholders, including employees, AI developers, and the organization, is essential.

Augmentation over Replacement: The insights from the MIT study reinforce the narrative that generative AI's role is to augment human capabilities rather than to replace human labor. It serves as a catalyst for skill enhancement, not a stand-in for human insight and decision-making.

Strategies for Integrating Generative AI into the Workforce

Tailored Training and Onboarding: Crafting comprehensive training programs that not only familiarize workers with generative AI tools but also deepen their understanding of the technology's recommendations is vital. Such education should emphasize the mechanics of AI tools and the logic behind their outputs, fostering a workforce that is both technologically adept and critically engaged.

Cultivating a Culture of Mentorship and Lifelong Learning: Encouraging a workplace ethos of mentorship and ongoing education can amplify the advantages of generative AI. Experienced professionals play a crucial role in guiding newer employees through the nuances of integrating AI into their workflows, ensuring that the technology serves as a bridge to greater understanding and skill rather than as a crutch.

Ensuring Ethical and Responsible AI Utilization: The deployment of generative AI comes with a responsibility to uphold ethical standards, particularly concerning privacy, data security, and the avoidance of over-dependence on technological solutions.

Generative AI stands as a transformative force poised to democratize access to skills and knowledge within the workforce, particularly empowering those at the early stages of their careers. By boosting productivity and expediting the learning process, generative AI has the potential to forge a more equitable, capable, and dynamically skilled workforce. However, this technological integration demands thoughtful management to ensure that it amplifies rather than eclipses human intellect, promoting an ecosystem where continuous growth and skill development are paramount.

AI-Native Thinking in Talent, Culture, and Organizational Structure

An AI-Native environment demands a workforce adept not only in technical AI skills but also in understanding its broader implications. This begins with strategic talent acquisition, where the focus is on recruiting individuals with a blend of AI, machine learning, and data science expertise, as well as soft skills like adaptability and creative problem-solving. Continuous skill development is crucial, with an emphasis on training programs that go beyond basic AI operation to include ethical AI use and collaborative problem-solving with AI systems.

The transition to an AI-centric culture is foundational in AI-Native thinking. This cultural shift requires nurturing an environment where AI is seen as a partner in innovation rather than a disruptor. Key to this transformation is cultivating openness to change, a continuous learning mindset, and adaptability among employees. Leadership must lead by example, advocating for and demonstrating the benefits of AI collaboration while also addressing ethical considerations and the responsible use of AI.

AI-Native organizations need to reevaluate their structures to support seamless AI integration. This might involve creating specialized AI and data management roles or departments, or embedding AI capabilities within various business units. Organizational structures should facilitate collaboration between AI systems and human employees, with clear guidelines for AI-informed decision-making processes. This structure must also be agile, capable of evolving with the rapidly changing AI landscape.

Effective change management is crucial in transitioning to an AI-Native business model. Employees' concerns about AI, such as job security and role transformation, must be addressed with transparency and empathy. Engaging employees in the AI transition process through clear communication, involvement in AI projects, and education on how AI can augment their roles is essential. This approach helps build a workforce that is not only skilled in AI but also supportive of its integration.

AI-Native thinking prioritizes ethical AI deployment. This includes establishing ethics guidelines, ensuring AI systems are free from biases, and promoting a diverse and inclusive workforce. Diversity in the team developing and managing AI systems is critical for mitigating biases in AI outputs. Ethical AI deployment also encompasses transparency in AI decisions, ensuring fairness and accountability in AI applications.

In an AI-Enhanced workplace, traditional job roles and performance management systems need to evolve. Job roles should be redesigned to focus more on tasks that AI cannot perform, such as creative and strategic thinking, complex problem-solving, and empathetic interactions. Performance management systems should assess the effectiveness of AI-human collaboration and recognize the evolving skill sets required in an AI-Native environment.

Employee well-being is a paramount concern in AI-Native thinking. Organizations must ensure that AI deployment does not lead to increased stress or unrealistic performance expectations. Ethical considerations regarding employee data privacy and AI-driven workforce analytics are critical. An AI-Native environment should support not just productivity and efficiency but also the mental and emotional health of its workforce.

In summary, AI-Native thinking in talent, culture, and organizational structure is about creating an ecosystem where AI integration is seamless, ethical, and enhances the human experience at work. By focusing on strategic talent acquisition, cultural transformation, organizational adaptation, effective change management, ethical AI deployment, job redesign, and employee well-being, organizations can fully harness the benefits of AI while fostering a resilient, adaptable, and innovative workforce.

The Future of Work with AI

The future of work with artificial intelligence (AI) presents a transformative shift in how we approach tasks, problem-solving, and innovation. This future is not characterized by AI replacing human workers, but rather by AI augmenting human abilities and opening new frontiers for collaboration and creativity.

Characteristics of the AI-Enhanced Workplace

- **Hybrid Human-AI Teams**: In the future workplace, we can expect hybrid teams where humans and AI systems work seamlessly together. AI will take on roles that maximize its strengths, such as data analysis and automation of routine tasks, while humans focus on areas that require emotional intelligence, moral judgment, and creative thinking.

- **New Job Roles and Opportunities**: The integration of AI in the workplace will lead to the creation of new job roles. These roles will focus on managing AI systems, interpreting AI-driven insights, and ensuring the ethical application of AI. This evolution will necessitate a workforce that is agile and adaptable to new technologies.

- **Workplace Automation and Upskilling**: While AI will automate certain tasks, this will not necessarily lead to reduced employment. Instead, it will shift the nature of work, emphasizing the need for upskilling and reskilling. Continuous learning will be a cornerstone of the AI-ready workforce, ensuring employees remain valuable and versatile in an AI-driven economy.

- **Democratization of Expertise**: AI tools will enable a broader range of employees to access and interpret complex data, effectively democratizing expertise. This access will empower more employees to make informed decisions and contribute to areas previously reserved for specialists.

Preparing for the Transition

- **Education and Training Systems**: Educational institutions and workplaces must evolve to prepare individuals for the AI-enhanced future. This includes incorporating AI education into curricula, offering training programs focused on AI and its applications, and fostering a culture of lifelong learning.

- **Policy and Governance**: As AI reshapes the workplace, policymakers will need to address issues such as workforce displacement, privacy concerns, and ethical use of AI. Effective governance will ensure that the benefits of AI are maximized while mitigating potential risks.

- **Mental and Emotional Adaptation**: The transition to an AI-driven workplace will require not just skill adaptation but also mental and emotional readiness. Employees will need to cultivate a mindset open to change, innovation, and continuous learning.

Leveraging AI for a Better Workplace

- **Enhanced Productivity and Creativity**: AI's ability to handle tedious tasks will allow employees to focus on creative and strategic activities, leading to higher job satisfaction and productivity.

- **Data-Driven Decision Making**: With AI's proficiency in data analysis, decision-making in the workplace will become more informed and evidence-based, reducing risks and enhancing efficiency.

- **Work-Life Balance and Flexibility**: AI can also contribute to better work-life balance, automating tasks that would otherwise extend working hours and enabling more flexible work arrangements.

The future of work with AI is not a dystopian landscape of joblessness, but a promising horizon of enhanced human-machine collaboration. This future will be marked by new opportunities, roles, and ways of working, requiring a workforce that is adaptable, skilled, and ready to embrace the AI revolution. By preparing for these changes, we can harness the full potential of AI to create a more efficient, innovative, and fulfilling work environment.

Chapter 12

AI Governance Excellence: The Responsible AI Framework and Program

"Good governance requires working toward common ground. It isn't easy."
—— *Pete Hoekstra*

I n an era where artificial intelligence (AI) transcends science fiction to become an integral part of our daily lives, how do we ensure that this powerful technology serves the greater good? What measures can organizations take to harness AI's potential while navigating the ethical minefields it presents? These questions lie at the heart of Responsible AI (RAI), a concept that has become increasingly crucial as AI systems play a larger role in decision-making processes, from healthcare diagnostics to financial lending.

The journey towards integrating RAI into organizational practices is both a challenge and an opportunity. It requires a delicate balance between leveraging AI for its immense capabilities and ensuring that its use aligns with ethical principles, societal values, and regulatory requirements. This chapter delves into the multifaceted approach needed to embed RAI deeply into the organizational fabric, guiding readers through the complexities of developing, implementing, and maintaining AI systems responsibly.

As we embark on this exploration, we confront the inherent tensions between innovation and ethics, efficiency and equity, and autonomy and control. How do organizations cultivate a culture that prioritizes ethical considerations without stifling innovation? What frameworks and strategies can be employed to ensure AI applications are fair, transparent, and accountable? And importantly, how can this journey towards responsible AI be sustained in an ever-changing technological landscape?

These are not mere rhetorical questions but pressing challenges that organizations face today. Addressing them requires a comprehensive understanding of RAI principles, a commitment to ethical AI use, and a proactive stance on continuous learning and adaptation. By examining the roles of leadership, policy integration, stakeholder engagement, and regulatory compliance, this chapter aims to equip organizations with the knowledge and tools needed to navigate the RAI landscape successfully.

Join us as we navigate the intricate path of Responsible AI, exploring how organizations can not only comply with ethical standards but also champion the advancement of AI in a way that benefits all of society. Through this journey, we uncover the strategies, challenges, and triumphs of integrating RAI into the core of organizational culture, ensuring that AI serves as a force for positive transformation in our increasingly digital world.

The Responsible AI (RAI): Core Principles and Frameworks

In the rapidly evolving landscape of Artificial Intelligence (AI), the incorporation of Responsible AI (RAI) practices within organizational frameworks has become crucial. As AI technologies burgeon, leading consultancy and advisory entities such as Gartner, McKinsey, and Deloitte have pioneered distinct frameworks to shepherd organizations in weaving RAI into their fabric. While each model presents a unique lens through which RAI can be viewed and implemented, they converge on several foundational pillars that underscore the essence of RAI.

Gartner's Framework delineates a comprehensive approach that spans from establishing robust data foundations to crafting strategies and planning, through development and execution, culminating in value generation. It accentuates a harmonious balance between the technical prowess and societal impacts of AI, advocating for a synthesis that nurtures both dimensions.

McKinsey's Model champions a holistic perspective on AI, integrating both its capabilities and the attendant risks. It foregrounds the integration of ethical AI into the organizational ethos, advocating for the operationalization of AI ethics through well-structured governance, risk management, and scalable practices that ensure ethical considerations are embedded in every facet of AI deployment.

Deloitte's Approach emphasizes the criticality of ethical technology deployment, with a keen focus on upholding ethical AI principles, instituting robust governance frameworks, and maintaining a human-centric orientation. Leadership's role in cultivating an ethical AI culture stands out as a pivotal element in their framework, signifying the strategic importance of top-down commitment to ethical AI.

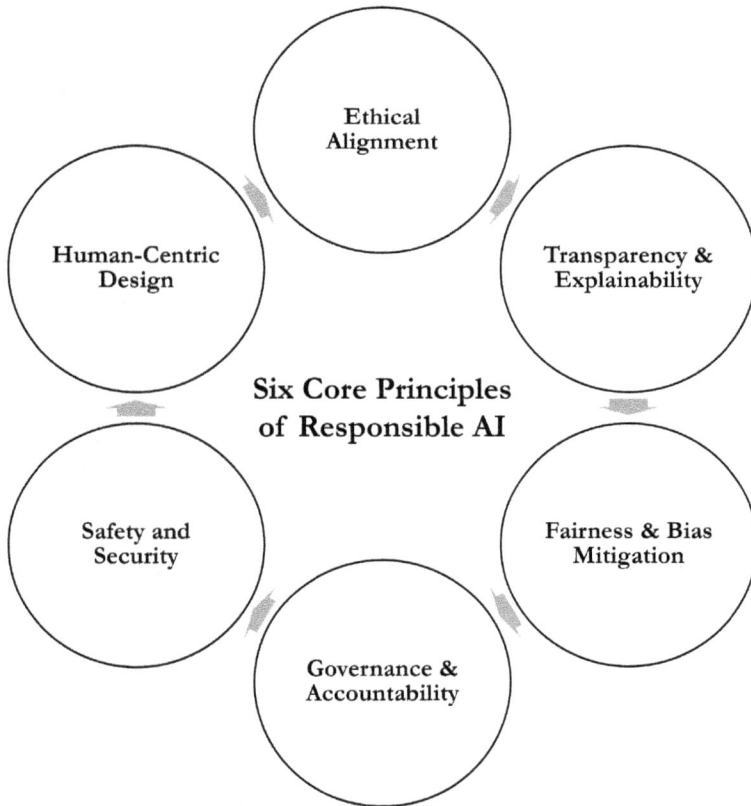

Figure 12-1: The Six Core Principles of Responsible AI (RAI)

Despite the diversity in their approaches, these frameworks coalesce around six core tenets that form the bedrock of Responsible AI:

- **Ethical Alignment:** The frameworks unanimously underscore the necessity of ensuring AI's alignment with ethical values and principles. This alignment ensures that AI's decision-making processes are congruent with human rights and societal norms, embodying a commitment to ethical stewardship.

- **Transparency and Explainability:** They advocate for AI systems that are transparent and explainable, fostering an environment where users and stakeholders can understand the mechanics behind AI decision-making. This transparency is vital for trust and accountability.

- **Fairness and Bias Mitigation:** A universal theme across the frameworks is the imperative to address biases and champion fairness within AI algorithms. This commitment is aimed at forestalling the perpetuation or emergence of systemic disadvantages, ensuring AI serves as a force for equity.

- **Governance and Accountability:** The establishment of robust governance structures and the accountability of AI systems and their creators are deemed indispensable. This includes ensuring mechanisms are in place for recourse and redress for those adversely affected by AI.

- **Safety and Security:** The frameworks collectively prioritize the safety and security of AI systems, emphasizing the need to safeguard against misuse and protect data privacy and integrity.

- **Human-Centric Design:** A shared emphasis on human-centric design underscores the importance of AI as a tool to augment human capabilities rather than replace them, ensuring that AI serves to enhance human endeavors rather than diminish them.

While Gartner, McKinsey, and Deloitte each offer distinct pathways to integrating RAI into organizational practices, their frameworks collectively illuminate the fundamental principles that should guide the development and deployment of AI technologies. These principles not only ensure the ethical, transparent, and equitable use of AI but also underscore the importance of human-centric design, safety, and governance as pillars of Responsible AI.

Crafting a Customized Approach to Responsible AI

Implementing Responsible AI (RAI) necessitates a bespoke strategy, acknowledging the diversity of organizational needs and contexts. A uniform solution is impractical; instead, organizations must sculpt their RAI strategies to fit their distinct objectives, ethical commitments, and societal responsibilities. This tailored approach ensures that AI initiatives resonate with an organization's specific business aims, ethical benchmarks, and broader societal duties.

The cornerstone of a tailored RAI strategy is a deep understanding of an organization's specific milieu, encompassing its industry landscape, regulatory framework, customer demographics, and technological prowess. This foundational knowledge is crucial for pinpointing the RAI facets most pertinent to the organization, facilitating a focused and effective implementation.

Step-by-Step Application of RAI Principles

1. **Ethical Alignment:** The application of ethical alignment in AI involves several critical steps. Firstly, it's essential to conduct an ethical audit of both current and planned AI systems. This process evaluates how these systems align with ethical standards and societal values. Following this, organizations should develop clear, comprehensive policies that delineate the ethical boundaries and expectations for AI usage. These policies must be communicated effectively within the organization. Additionally, a crucial step is the education and training of both AI developers and users in these ethical considerations, ensuring that everyone involved in the AI lifecycle is aware of and committed to upholding these standards.

2. **Transparency and Explainability:** For transparency and explainability, organizations should focus on the design of AI systems. This includes creating user-friendly interfaces that can elucidate AI decisions in terms that are easy to understand for all stakeholders. Moreover, there is a need for ongoing communication about how AI systems function. This involves explaining the decision-making processes of AI systems to stakeholders in a clear and comprehensible manner, which helps in building trust and understanding.

3. **Fairness and Bias Prevention:** To ensure fairness and prevent bias, continuous analysis and monitoring of data used by AI systems are imperative. This involves scrutinizing the data for any inherent biases that might skew AI decisions. Alongside data analysis, regular testing of algorithms is necessary to ensure they are fair and produce unbiased outcomes. These tests should be part of a routine process to identify and rectify any issues of bias or unfairness in AI operations.

4. **Governance and Accountability:** Establishing a robust governance structure is a fundamental step in ensuring responsible AI usage. This structure should oversee the use of AI across the organization and ensure adherence to ethical standards and practices. Equally important is the creation of mechanisms for accountability and recourse in instances where AI systems cause harm or errors. These mechanisms ensure that there are clear paths for addressing any negative impacts of AI and for holding the appropriate entities responsible.

5. **Safety and Security:** The safety and security of AI systems are paramount. Implementing strong security measures to protect AI systems

from breaches and unauthorized access is a critical step. Alongside security, regular safety tests and risk assessments of AI systems are necessary to ensure they operate safely and as intended, minimizing risks to users and other stakeholders.

6. **Human-Centric Design:** A human-centric approach to AI involves engaging end-users in the AI system design process. This engagement ensures that the systems are tailored to real user needs and are accessible and understandable. Additionally, assessing the potential impacts of AI on human jobs and well-being is crucial. This assessment should consider how AI will affect employment, workflows, and the psychological impact on employees and users, ensuring that the deployment of AI enhances human capabilities rather than undermining them.

In summary, a nuanced, context-aware approach to RAI, grounded in a firm understanding of an organization's unique environment and augmented by a committed application of core RAI principles, is essential for fostering ethical, transparent, and equitable AI solutions. This strategy not only aligns AI initiatives with organizational values and societal expectations but also ensures that AI serves as a beneficial, augmentative force within the organization.

Customizing Responsible AI Strategies Across Industries

Implementing Responsible AI (RAI) across diverse industries necessitates a nuanced approach, aligning with the unique challenges and operational landscapes of each sector. Below are some examples showcasing the adaptation of RAI principles within various industry contexts, emphasizing the importance of a tailored strategy to meet specific needs.

Financial Institutions: Ensuring Equity in AI-Driven Credit Scoring

The use of AI in credit scoring presents a significant challenge for financial institutions, particularly the risk of unintentional discrimination against certain groups. To counteract this, these institutions are adopting a multi-faceted RAI approach. This includes the integration of advanced analytics to detect and correct biases continuously, coupled with the development of tools designed to clarify credit decisions for applicants, thereby enhancing transparency. Furthermore, engaging community representatives to review and provide feedback on these models fosters a more inclusive financial ecosystem.

Healthcare Providers: Enhancing Trust with Transparent AI Diagnostics

In healthcare, the opacity of AI-driven diagnostic tools can undermine trust among clinicians and patients. Addressing this challenge, healthcare providers are prioritizing the deployment of explainable AI systems that offer comprehensible justifications for their recommendations. This effort is complemented by initiatives to educate healthcare professionals on interpreting AI diagnostics and communicating them to patients effectively. Moreover, the design of patient-centric AI interfaces aims to demystify diagnostic processes, empowering patients with clear, accessible information about their health outcomes.

E-Commerce Platforms: Navigating Personalization and Privacy

E-commerce platforms leverage AI to offer personalized shopping experiences, a strategy that walks a fine line between enhancing user engagement and encroaching on customer privacy. To balance these considerations, platforms are implementing stringent data privacy measures and conducting regular audits of recommendation algorithms to uncover and address biases. This ensures a diverse and equitable exposure to product offerings. Additionally, by empowering customers with greater control over their data, e-commerce platforms bolster consumer trust and security, reinforcing the ethical use of AI in personalization efforts.

Each sector's adaptation of RAI principles illustrates a deep understanding of industry-specific challenges and the strategic application of RAI to mitigate these issues. This tailored application not only aligns with ethical standards but also builds a foundation for trust, efficiency, and sustainability in AI initiatives. By customizing RAI approaches, industries can navigate the complex interplay between technological innovation and ethical responsibility, ensuring that AI serves as a force for good in their operational landscapes.

Cultivating AI Governance Excellence through a Comprehensive RAI Program

The development of a robust Responsible AI (RAI) program is essential for achieving AI governance excellence. This program should be comprehensive, adaptable, and aligned with both the organization's objectives and evolving industry best practices, many of which draw parallels to established cybersecurity programs.

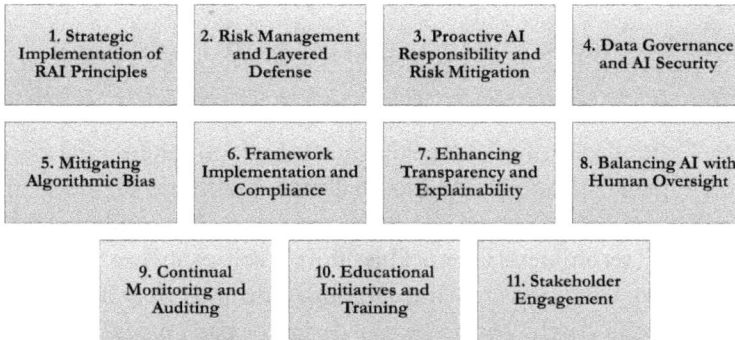

1. Strategic Implementation of RAI Principles	2. Risk Management and Layered Defense	3. Proactive AI Responsibility and Risk Mitigation	4. Data Governance and AI Security
5. Mitigating Algorithmic Bias	6. Framework Implementation and Compliance	7. Enhancing Transparency and Explainability	8. Balancing AI with Human Oversight
9. Continual Monitoring and Auditing	10. Educational Initiatives and Training	11. Stakeholder Engagement	

Figure 12-2: The 11 Core Components of the RAI Program

Key Components of a Robust RAI Program

1. **Strategic Implementation of RAI Principles**: The program must go beyond theoretical principles, embedding them into the organization's operations through written protocols and procedures. This strategic implementation involves mapping each RAI principle to specific guidelines, policies, and procedures, ensuring these principles are actively practiced across the organization.

2. **Risk Management and Layered Defense**: Incorporating a multi-layered defense strategy is crucial. This involves mobilizing different teams responsible for specific risks at various stages of AI development and deployment. Each team contributes to managing risks and reinforcing responsible AI behaviors, echoing the multi-faceted approach seen in cybersecurity programs.

3. **Proactive AI Responsibility and Risk Mitigation**: Clear accountability must be established for the development and deployment of AI systems. This includes implementing transparent documentation, bias detection, and fairness mechanisms, as well as advanced techniques for addressing potential risks such as filter bubbles and biased reinforcement.

4. **Data Governance and AI Security**: Robust data governance and security are paramount. This involves developing AI-specific data governance rules and security standards to mitigate risks associated with data misuse and to uphold the integrity of AI outcomes.

5. **Mitigating Algorithmic Bias**: Unintended biases in AI systems can lead to unfair practices. Addressing this challenge involves employing pre- and

post-processing techniques and tools for identifying and rectifying biases, ensuring AI systems operate without unjust discrimination.

6. **Framework Implementation and Compliance**: Implementing AI governance frameworks with robust safeguards ensures compliance with regulatory standards. This includes establishing a reporting structure to senior leadership, fostering an organizational culture that prioritizes AI ethics, and conducting regular audits.

7. **Enhancing Transparency and Explainability**: Improving model transparency is crucial for accountability in automated decision-making. Developing understandable AI systems with clear documentation enhances trust and allows for scrutiny of AI operations.

8. **Balancing AI with Human Oversight**: Integrating human judgment in key AI decision-making processes, combined with ongoing monitoring, ensures a balance between AI efficiency and ethical responsibility. This approach is akin to the 'human-in-the-loop' concept in cybersecurity.

9. **Continual Monitoring and Auditing**: Regular monitoring and auditing of AI systems are essential for detecting issues and ensuring compliance with ethical standards and organizational objectives.

10. **Educational Initiatives and Training**: Ongoing education and training for staff are vital. Keeping teams updated on ethical considerations, industry best practices, and trends in AI technology ensures responsible navigation of the AI landscape.

11. **Stakeholder Engagement**: Involving customers, users, and the public in AI-related decisions fosters transparency, trust, and accountability. This engagement ensures that AI applications align with societal expectations and values.

Establishing a robust RAI program is integral to achieving AI governance excellence. By incorporating these components, organizations can ensure their AI initiatives are not only technologically advanced but also ethically sound, legally compliant, and socially responsible. This approach aligns with the growing recognition of the need for comprehensive oversight in the AI domain, paralleling the rigor and depth of cybersecurity programs.

Evaluating Responsible AI: Metrics for Success

The foundation of a successful Responsible AI (RAI) program lies in the ability to measure and understand the impact of AI initiatives accurately. Instituting a standard set of quantifiable metrics, tailored to an organization's unique business context, is crucial. These metrics provide a systematic approach to track various dimensions such as AI model quality, usage levels across different business units, measurable impacts based on the purposes and intended users, and long-term outcomes. Tailoring these metrics ensures they are directly relevant to the specific goals and challenges of the organization, enabling a more precise evaluation of AI's effectiveness and value.

Crafting Custom Metrics for Comprehensive Insight

- **AI Model Quality**: This involves metrics that assess the accuracy, reliability, and efficiency of AI models. It could include error rates, predictive accuracy, and the degree of bias present in the models. Tailoring these metrics to the organization's specific AI applications ensures they are relevant and meaningful.

- **Usage Levels Across Business Units**: Tracking how different departments or business units utilize AI helps in understanding adoption rates and identifying areas where AI can be more effectively leveraged. Metrics might include the number of AI-driven processes within each unit or the frequency of AI system usage.

- **Measurable Impact Metrics Based on Purpose and Users**: These metrics should be designed to assess the impact of AI systems on their intended users. For example, in customer service applications, metrics could include customer satisfaction scores or reduction in response times.

- **Long-Term Business Outcomes**: It's important to measure the long-term effects of AI on business outcomes. This could involve metrics like ROI, customer retention rates, or overall efficiency improvements over time.

Gathering Metrics Through a Centralized Governance Platform

A unified governance platform for AI offers comprehensive insights into AI's performance, system health, and business impact, ensuring:

- **Enterprise-Wide Visibility**: A centralized governance platform for AI can provide a holistic view of AI adoption levels, system health, and the business value delivered by AI across the entire organization. This approach facilitates a unified view of AI performance and its impact on various aspects of the business.

- **System Health Monitoring**: Such a platform should include tools to continuously monitor the health and performance of AI systems. This can involve real-time diagnostics, anomaly detection, and alerting mechanisms for potential issues.

- **Business Value Analysis**: The platform should be capable of analyzing and presenting the business value derived from AI initiatives. This includes quantifiable benefits like cost savings, as well as qualitative benefits like improved customer engagement or employee satisfaction.

- **Data Integration and Accessibility**: For effective governance, the platform must integrate data from various sources within the organization, ensuring that the metrics are comprehensive and up-to-date. It should also be accessible to relevant stakeholders, enabling informed decision-making.

- **Compliance and Ethical Standards Monitoring**: The platform can also be used to ensure that AI systems are compliant with relevant laws and regulations and adhere to ethical standards. This involves monitoring for compliance with data privacy laws, ethical AI guidelines, and industry-specific regulations.

Implementing a central AI governance platform equipped with tailored, measurable metrics is crucial for tracking and understanding the effectiveness and impact of AI systems. This strategic approach not only sheds light on AI's immediate benefits but also aligns AI efforts with broader business goals and ethical considerations, fostering a responsible and value-driven AI landscape.

AI-Native Thinking: Revolutionizing Responsible AI Governance

AI-native thinking is a transformative approach that redefines Responsible AI (RAI) governance by embedding AI insights and considerations at the core of organizational governance structures. This forward-thinking perspective ensures that AI is not merely an add-on but a fundamental aspect of decision-making, strategic planning, and operational processes. It requires a holistic and integrated approach to governance that anticipates the unique challenges and opportunities presented by AI technologies.

- **Strategic AI Integration**: AI-native governance involves integrating AI considerations into the very DNA of organizational strategy. This means recognizing AI's potential to redefine business models, operational efficiencies, and market strategies from the outset, ensuring AI governance is proactive rather than reactive.

- **Proactive Ethical Management**: Central to AI-native thinking is the anticipation and preemptive resolution of ethical dilemmas posed by AI applications. This approach embeds ethical considerations into the life-cycle of AI systems, from conception through deployment, ensuring that ethical governance is a continuous process.

- **Data-Centric Governance**: At the heart of AI-native governance is a data-centric approach that prioritizes the ethical collection, use, and management of data. This ensures that data practices not only comply with regulatory standards but also align with broader ethical principles, enhancing the integrity and trustworthiness of AI systems.

- **Dynamic Policy Frameworks**: AI-native thinking demands governance frameworks that are as dynamic and adaptable as the AI technologies they seek to regulate. This involves creating policies that can evolve in response to new developments in AI, ensuring governance mechanisms remain relevant and effective.

- **Stakeholder Engagement in AI Development**: Emphasizing collaboration and transparency, AI-native governance incorporates diverse stakeholder perspectives into the AI development process. This inclusive approach fosters broader acceptance and trust in AI applications, ensuring they are responsive to the needs and concerns of all affected parties.

- **Continuous Learning and Adaptation**: Recognizing the rapid pace of AI innovation, AI-native governance embraces continual learning and adaptation. This commitment to ongoing education and policy evolution ensures that governance practices keep pace with technological advancements, maintaining their relevance and effectiveness.

- **Cultivation of an AI-Savvy Culture**: AI-native governance also extends to cultivating an organizational culture that understands and appreciates the implications of AI. This cultural shift ensures that all levels of the organization are prepared to engage with AI responsibly and effectively.

By adopting an AI-native approach to governance, organizations can ensure their RAI practices are not only compliant with current standards but are also positioned to adapt to future challenges. AI-native thinking empowers organizations to harness the full potential of AI technologies in a responsible, ethical, and sustainable manner. It signifies a shift towards governance models that are inherently flexible, ethically grounded, and aligned with the transformative potential of AI, setting a new standard for excellence in RAI governance.

Fostering a Culture of Responsible AI within Organizations

Integrating Responsible AI (RAI) into an organization's culture is a nuanced endeavor that extends beyond mere compliance with guidelines. It involves infusing RAI values into the very essence of the organization, influencing decision-making, strategy development, and daily operations. This comprehensive approach ensures the ethical use of AI, aligning with societal values to foster a foundation of responsibility, innovation, and trust.

The shift towards a culture imbued with RAI principles starts with leadership. Executives must not only advocate for RAI but also demonstrate their commitment through their actions, setting a vision for RAI that aligns with the organization's overall mission. This vision must be clearly communicated and embraced at all organizational levels, ensuring a unified understanding and commitment to RAI principles.

RAI principles should be reflected across all organizational policies and practices. This means embedding RAI considerations into HR guidelines, employee training programs, product development processes, and customer engagement strategies. For instance, HR policies could promote ethical AI usage, while product development procedures might include RAI checkpoints to scrutinize AI applications at each development stage.

Education and awareness-building among all employees are crucial for creating an RAI-centric culture. This includes regular training sessions and workshops to keep staff informed about the latest RAI principles and practices. Such education should reach beyond technical teams, encompassing all departments to highlight RAI's multidisciplinary impact.

Fostering an environment that encourages open dialogue about AI's ethical implications is essential. Employees should feel empowered to voice concerns and propose improvements regarding AI applications. Additionally, engaging with external stakeholders like customers, regulators, and the community can offer diverse perspectives, enriching the organization's RAI approach.

To truly embed RAI into the organizational culture, it's vital to implement mechanisms for measuring and reporting on RAI progress. This could involve establishing specific RAI goals, monitoring performance against these objectives, and publicly sharing outcomes and initiatives. Such transparency not only holds the organization accountable but also signals its commitment to RAI to external stakeholders.

Finally, recognizing and rewarding efforts that advance RAI within the organization can underscore its importance. Formal recognition programs, incentives, or incorporating RAI considerations into performance evaluations can acknowledge the contributions of teams and individuals, reinforcing the organization's dedication to responsible AI practices.

By adhering to these principles, organizations can ensure their AI initiatives are not just technologically advanced but also ethically responsible and socially aligned, creating a culture that values innovation and responsibility in equal measure.

Responsible AI as a Continuous Journey

Recognizing RAI as a continuous journey underscores the importance of adaptability, vigilance, and ongoing engagement. It's a commitment to perpetual learning, evolution, and improvement in the face of an ever-changing technological and societal landscape. In this journey, organizations not only contribute to the ethical advancement of AI but also position themselves as resilient and forward-thinking in a rapidly evolving digital world.

Understanding Responsible AI (RAI) as a continuous journey is pivotal for organizations aiming to integrate AI ethically and effectively. Unlike static compliance targets, RAI is dynamic, evolving with technological advancements, societal values, and regulatory landscapes. This evolutionary approach acknowledges that as our understanding of AI's impact deepens, the strategies and practices surrounding its use must also adapt and grow.

AI technology is advancing at a rapid pace, bringing new capabilities and complexities. As these technologies evolve, so too do the challenges and opportunities associated with their responsible use. Organizations must stay abreast of the latest developments in AI and continually reassess their RAI strategies to ensure they remain relevant and effective. This includes adopting new tools and methodologies for fairness testing, privacy preservation, and security, as well as updating policies and practices to address emerging issues.

Societal norms and values are not static, and neither are the expectations placed on AI systems. What is considered ethical and responsible today may change as societal attitudes evolve. Organizations need to be sensitive to these shifts and be prepared to adjust their RAI approaches accordingly. This requires ongoing dialogue with stakeholders, including customers, employees, regulators, and the wider community, to understand their perspectives and concerns.

The regulatory landscape for AI is also in a state of flux. As governments and international bodies grapple with the implications of AI, new regulations and guidelines are being introduced. Organizations must be agile in responding to these changes, ensuring that their AI systems and practices remain compliant with current and future regulations. This involves not just following the letter of the law but also understanding its spirit and intent.

A key aspect of viewing RAI as a continuous journey is fostering a culture of continuous learning and improvement within the organization. This means encouraging experimentation and innovation while also being open to learning from mistakes. It involves regular reviews of RAI initiatives, soliciting feedback, and making iterative improvements.

Effective RAI implementation requires mechanisms for capturing feedback and integrating learning back into AI systems and practices. This could involve user feedback systems, AI system audits, or post-deployment reviews. By creating feedback loops, organizations can learn from real-world experiences and refine their RAI approaches over time.

Finally, staying informed about the broader RAI discourse is crucial. This involves engaging with industry groups, participating in forums and conferences, and collaborating with researchers and practitioners in the field. By staying engaged with the wider RAI community, organizations can learn from others' experiences and contribute to the collective understanding of responsible AI.

Chapter 13

AI Technology Foundation: A Comprehensive Guide

"The great growling engine of change – technology." —— *Alvin Toffler*

How do we transform the abstract potential of Artificial Intelligence (AI) into tangible outcomes that redefine the landscape of business and technology? What foundational elements must be in place for AI to not just thrive but revolutionize industries, innovation, and our very approach to problem-solving? These questions lie at the heart of the AI Technology Foundation, a complex and indispensable framework that supports the successful execution of AI strategies. This chapter is an invitation to explore the depth and breadth of this foundation, uncovering the critical layers and components that enable AI's transformative power.

At the forefront of our exploration is the AI Technology Stack, a framework meticulously organized into four interconnected layers—Application, Platform, Model, and Infrastructure. Each layer serves a distinct purpose, collectively ensuring that organizations are well-equipped to support their AI initiatives from inception to deployment. But how are these layers structured, and what unique role does each play in the lifecycle of AI implementation?

The journey into generative AI deployment unveils another layer of complexity and innovation. Through the lens of foundation models in vision, speech, and language, we witness a leap forward in AI development. But what challenges and opportunities do these models present? Deployment strategies such as Model Training, Fine-Tuning, Retrieval-Augmented Generation (RAG), and Prompt Engineering offer tailored solutions, but selecting the right strategy is key to unlocking the full potential of AI tailored to organizational needs.

Choosing the right AI tools emerges as a critical decision point, influenced by factors such as prototype development time, scalability, collaboration capabilities, and compliance adherence. But what makes a tool fit for purpose in the rapidly evolving AI landscape? And how do these tools contribute to the effective management and monitoring of AI applications?

Diving deeper, the realms of MLOps, AIOps, and AI Lifecycle Management reveal the essential disciplines for the successful deployment and ongoing management of AI systems. These areas reflect the rapid pace of AI innovation, focusing on operational efficiency, scalability, robustness, and transparency. But what challenges do organizations face in implementing these disciplines, and how can they navigate the complexities of the AI lifecycle?

This chapter aims to not just inform but engage you in the intricacies of the AI Technology Foundation. Through this comprehensive guide, we delve into the confluence of technology, strategy, and people, emphasizing the critical role of this foundation in empowering organizations to harness AI's power effectively. The journey through the AI Technology Foundation is a journey of discovery, under-standing, and appreciation for the components that come together to support and drive AI initiatives towards achieving transformative business objectives.

AI Technology Stack

The AI technology stack serves as the foundational framework that enables the seamless execution of Artificial Intelligence (AI) from the initial stages of data acquisition through to the deployment of sophisticated AI-based solutions. This intricately designed stack is divided into four critical, interconnected layers, with each layer playing a unique role in addressing the diverse aspects of AI technology:

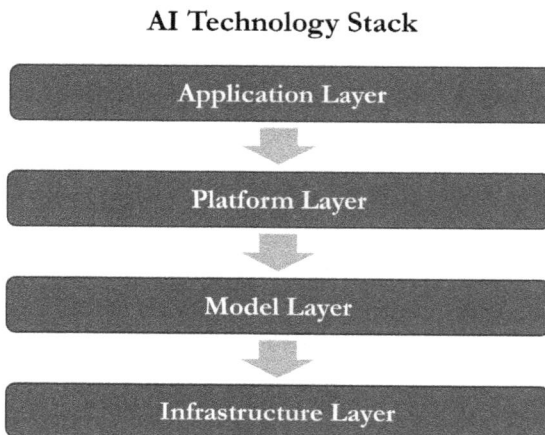

Figure 13-1: The Four Layers of AI Technology Stack

1. **Application Layer:** This is where AI meets the end user, facilitating direct interaction between humans and AI systems. It's the layer where the dynamic generation of content comes to life, driven by specialized algorithms that tailor user experiences. Here, applications leverage AI to deliver personalized content, make recommendations, and enable natural language interactions, effectively bridging the gap between complex AI technologies and user-friendly interfaces.

2. **Platform Layer:** Acting as the gateway to Large Language Models (LLMs), the platform layer democratizes access to some of the most advanced AI capabilities available today. It encompasses managed services and platforms that offer seamless integration with pre-trained models, making it easier for developers to fine-tune and customize these models to fit specific requirements. This layer includes both proprietary cloud platforms like Azure OpenAI Service, Amazon Bedrock, and Google Cloud's Vertex AI, as well as community-driven, open-source platforms such as Hugging Face, TensorFlow, and PyTorch, each offering unique tools and resources for AI development.

3. **Model Layer:** Positioned at the core of generative AI (GenAI), the model layer is where the magic begins with foundation models. These models, based on groundbreaking Transformer algorithms, are trained on vast amounts of unlabeled data, making them incredibly versatile for a wide array of tasks. The accessibility of these models varies, with closed source models like GPT-4 and PaLM 2 being available through APIs, offering ease of use but less transparency. In contrast, open-source models provide a platform for greater customization and insight into the model's inner workings, fostering innovation and adaptability.

4. **Infrastructure Layer:** The unsung hero of GenAI, the infrastructure layer, provides the necessary backbone to support both the training and inference phases of AI models. This includes the hardware and software required for the heavy lifting behind the scenes – from semiconductors, networking, and storage solutions to databases and cloud services. It's here that the computational power of GPUs, TPUs, and FPGAs becomes essential, enabling the rapid processing and analysis of data that drives AI innovation forward.

Together, these layers form a cohesive ecosystem that supports the development and deployment of generative AI across a variety of domains. Each layer contributes to the seamless operation of AI technologies, ensuring that from the moment a user interacts with an AI application to the underlying computational processes, the AI technology stack stands as the pivotal foundation for AI-driven solutions.

AI Application Layer: Bridging AI and User Experience

The Application Layer in Generative AI architecture is pivotal in making artificial intelligence not only accessible but also valuable to a broad spectrum of users. This layer serves as the critical interface that bridges the gap between the intricate workings of AI models and the practical needs of end-users, facilitating the dynamic generation and application of AI-created content. Here, we delve into the multifaceted aspects of the Application Layer, highlighting its significance in the realm of generative AI.

User Interface and Interaction

At the forefront of the Application Layer is the user interface (UI), the platform where direct interaction between users and AI models occurs. These interfaces are crafted to be intuitive, allowing users to effortlessly input data, customize parameters, and receive AI-generated outputs. For example, a chatbot interface leveraging a language model enables users to submit text prompts and receive coherent, contextually relevant responses, demonstrating the seamless integration of AI into user-friendly platforms.

Types of Applications

- **Generalized Applications**: Designed with versatility in mind, these applications cater to a broad spectrum of tasks across various domains, from generating text and images to creating videos, audio, and software code. ChatGPT and DALL-E 2 stand as exemplars, showcasing the wide-ranging capabilities of generative AI in producing diverse content forms.

- **Domain-Specific Applications**: These applications are finely tuned to meet the distinct needs of specific industries such as finance, healthcare, or education, leveraging specialized datasets to deliver targeted solutions. BloombergGPT and Google's Med-PaLM 2 are prime examples, offering advanced capabilities in financial data analysis and medical query responses, respectively.

- **Integrated Applications**: This category encompasses traditional software solutions that have embraced generative AI functionalities to augment their core features. Through such integration, AI capabilities are woven into the fabric of existing software ecosystems, enhancing productivity and creativity. Noteworthy integrations include Microsoft 365's AI-enhanced productivity tools, Salesforce CRM's Einstein GPT for customer relationship management, and Adobe Creative Cloud's adoption of generative AI in creative workflows.

Development and Customization

The Application Layer also extends to the creation of new AI-powered applications by both foundational model proprietors and third-party developers. This process often involves utilizing open-source frameworks and proprietary models to forge applications tailored to specific tasks or demographics, emphasizing the layer's role in fostering innovation and customization.

User Experience and Accessibility

A cornerstone of the Application Layer is its emphasis on user experience (UX) design, ensuring that AI applications are not only intuitive but also accessible to a wide audience. This commitment to UX design aims to make sophisticated AI technologies approachable and usable for individuals regardless of their technical expertise.

Real-world Implementations

In practical terms, the Application Layer is the stage where the theoretical potential of generative AI materializes into concrete, user-centered applications. From personalizing content and automating mundane tasks to enhancing creative endeavors and providing insightful analytics, this layer translates AI's capabilities into solutions that tackle real-world challenges and needs.

In essence, the Application Layer of Generative AI is a testament to how advanced AI models are being transformed into practical, accessible applications. By democratizing AI technology and fostering innovation across sectors, this layer plays a crucial role in translating complex AI capabilities into user-centric solutions that address a wide array of challenges and opportunities.

AI Platform Layer: The Bridge to Advanced AI Capabilities

The Platform Layer stands as a cornerstone within the Generative AI ecosystem, seamlessly connecting the dots between foundational AI models and their real-world applications. This layer is instrumental in democratizing access to Large Language Models (LLMs) and streamlining their integration into diverse use cases. It simplifies the otherwise complex process of adapting and customizing generative AI models, making advanced AI technologies accessible to a broader audience. Let's delve into the critical elements that define the Platform Layer's role in the AI landscape.

Access to Foundation Models

Central to the Platform Layer is its provision of access to pre-trained foundation models, such as OpenAI's GPT series. This accessibility is crucial for organizations and developers aiming to leverage cutting-edge AI capabilities without the substantial investment required to develop these models from the ground up.

Fine-Tuning and Customization

A key function of the Platform Layer is facilitating the fine-tuning of LLMs. This process involves retraining a pre-existing model on a specialized dataset to tailor it to specific tasks or domains, thereby enhancing the model's accuracy and effectiveness for targeted applications.

Managed Services and Simplification

The layer is characterized by managed services offered by Cloud Service Providers (CSPs), which alleviate the intricacies associated with infrastructure, model management, and scalability. These services empower companies to employ LLMs efficiently, bypassing the need for in-depth technical knowledge in AI development and maintenance.

Cloud Platforms for AI Model Fine-Tuning

CSPs have curated specialized services that provide foundational model access and support the bespoke training and customization of AI models, including:

- **Azure OpenAI Service**: Integrates OpenAI's models within the Azure ecosystem, facilitating text generation with GPT and code generation with Codex.

- **Amazon Bedrock**: Aims to streamline the development and scaling of generative AI applications, utilizing models like Anthropic's Claude and Stability AI's Stable Diffusion.

- **Google Cloud's Vertex AI**: Offers a comprehensive ML platform equipped with tools for AI model building, training, and deployment.

Open Source Platforms

The ecosystem is further enriched by open-source platforms, contributing significantly to the Platform Layer:

- **Hugging Face**: Acts as a central hub for pre-trained transformer models, alongside resources for model fine-tuning across various NLP tasks.

- **TensorFlow and PyTorch**: These leading open-source libraries provide extensive support for the development, training, and deployment of AI models, catering to a wide range of applications.

Cost Efficiency and Accessibility

By leveraging the services within the Platform Layer, organizations can markedly reduce the costs associated with the development and maintenance of generative AI models. These platforms often offer flexible subscription models or integrate their services with Infrastructure as a Service (IaaS) offerings, enhancing the accessibility of advanced AI capabilities.

Security and Privacy Features

Security and privacy are paramount within the Platform Layer, with services incorporating robust measures to ensure that data involved in model training and generated outputs are handled securely and in compliance with regulatory standards.

The Platform Layer in Generative AI is pivotal in broadening access to sophisticated AI technologies. By facilitating the streamlined use and customization of large-scale AI models, it enables a diverse array of users to explore and innovate with AI, driving forward a multitude of advanced applications.

AI Model Layer: The Engine of Generative AI

The Model Layer forms the heart of the Generative AI architecture, acting as the crucible where the foundational capabilities of AI are developed, refined, and expanded. This layer is responsible for crafting the sophisticated models that serve as the backbone for a multitude of generative AI applications. Let's explore the essential attributes and dynamics of the Model Layer in detail.

Foundation Models: The Core

At the center of the Model Layer lie the large-scale machine learning models, famously known as foundation models. These powerhouses, trained on vast, diverse datasets, possess the remarkable ability to understand and generate content across a spectrum of domains, such as language, vision, robotics, reasoning, and search. Iconic examples include OpenAI's GPT series, Google's BERT, and vision-centric models like DALL-E 2.

Training and Adaptation

Foundation models undergo extensive training, often through unsupervised learning techniques, allowing them to distill a broad comprehension of language, concepts, and patterns from massive pools of unlabeled data. This foundational knowledge enables subsequent fine-tuning or adaptation for specific tasks, significantly boosting their utility and precision for particular applications.

Closed Source vs. Open Source Models

The landscape of the Model Layer is diverse, encompassing both closed source models, developed and maintained by proprietary entities, and open source models, which are freely available for modification and use by the community. This dichotomy fosters a rich ecosystem of innovation and application, with closed source models like GPT-4 and Google's PaLM 2 offering streamlined access via APIs, and open source models encouraging deeper exploration and customization.

Model Training and Fine-Tuning

Training these behemoth models is a Herculean task, demanding vast computational resources to tweak millions of parameters. Fine-tuning, a subsequent stage, tailors these models to specialized datasets, honing their capabilities for distinct tasks or industries, a process pivotal for achieving high levels of accuracy and relevance.

Diversity of Foundation Models

The generative AI field thrives on a rich variety of foundation models, each bringing unique capabilities to the forefront of innovation in NLP, computer vision, and beyond. The anticipated growth in the reliance on these models, especially for NLP use cases by 2027, underscores their expanding influence and utility.

Domain-Specific Models

Beyond general-purpose models, the Model Layer also curates domain-specific models, meticulously refined from the broader foundation models to address the nuanced requirements of particular industries or tasks. These specialized models offer tailored solutions, enhancing effectiveness and precision.

Evolution and Innovation

The Model Layer is in a constant state of flux, with ongoing development of new models and the refinement of existing ones. It embodies the pinnacle of AI research and development, pushing the envelope of what AI can achieve.

Multimodal Models: A New Frontier

A notable advancement within the Model Layer is the emergence of multimodal models. These models are designed to interpret and generate data across multiple formats—text, images, audio, and video—simultaneously. They represent a leap towards more integrated AI systems capable of comprehensive understanding and interaction with the world.

- **Integration of Different Data Types**: Multimodal models excel in processing various data types in unison, enhancing their ability to comprehend context and content more holistically.

- **Training on Diverse Datasets**: By training on datasets that span multiple data formats, these models learn to correlate and interpret information from disparate sources, paving the way for innovative applications.

- **Layered Neural Networks**: Employing deep neural networks with specialized layers for different data types, multimodal models achieve a unified understanding of complex datasets.

- **Feature Extraction and Fusion**: Key elements from each data type are extracted and merged, allowing the model to make informed inferences based on a composite data view.

- **Real-World Applications**: From autonomous vehicles to interactive AI systems, multimodal models find utility in scenarios where diverse data inputs are natural and necessary.

Developing multimodal models introduces challenges, particularly in aligning and integrating varied data sources meaningfully. Additionally, the demand for extensive and diverse training datasets poses its own set of hurdles.

In essence, the Model Layer is the dynamic core of Generative AI, driving the field's advancements and broadening the horizons of AI's application in society. It stands as a testament to the ongoing journey of AI development, highlighting the blend of complexity, innovation, and aspiration that defines this exciting area of technology.

AI Infrastructure Layer: The Technological Bedrock of GenAI

The Infrastructure Layer forms the backbone of Generative AI architecture, delivering the crucial technological underpinnings essential for training, deploying, and operating large-scale AI models efficiently. This layer encompasses a broad spectrum of components and resources, each playing a pivotal role in the seamless functioning of AI systems. Let's navigate through the critical elements that constitute this foundational layer.

Hardware Infrastructure

Semiconductors: The linchpins of computation power for AI, semiconductors facilitate the production of GPUs (Graphics Processing Units), TPUs (Tensor Processing Units), and ASICs (Application-Specific Integrated Circuits), which are instrumental in handling AI workloads.

Accelerated Computing Hardware: GPUs have transcended their original role in graphics rendering to become indispensable in machine learning for their parallel computation capabilities. Platforms like NVIDIA's CUDA have been specifically developed to optimize GPU performance for AI tasks.

Specialized AI Processors: Devices such as TPUs, designed to accelerate Tensor-Flow operations, and NPUs (Neural Processing Units), underscore the industry's push towards hardware that is tailor-made for the demands of neural network processing.

Networking

Efficient data transfer mechanisms within AI systems, especially for server-to-server communication in computing clusters, are facilitated by advanced networking technologies like InfiniBand and high-bandwidth interconnects such as NVLink, ensuring swift and effective data handling for AI applications.

Storage Systems

Given the voluminous data requirements for AI model training and operation, parallel storage systems are employed to bolster data transfer rates, enabling rapid access to training datasets and efficient writing of model parameters.

Software Infrastructure

This includes a suite of machine learning frameworks and libraries, such as Tensor-Flow and PyTorch, which provide the essential tools for AI model development, training, and deployment, streamlining the creation of sophisticated AI applications.

Network Infrastructure

The backbone that supports the heavy bandwidth demands of AI workloads, network infrastructure is critical for the fluid transfer of large data volumes within and across AI systems, ensuring high performance and scalability.

Privacy and Security

With AI applications often dealing with sensitive information, implementing stringent security measures within the infrastructure is paramount to safeguard data integrity and ensure user privacy.

Cloud Services

Cloud platforms play a vital role in the AI ecosystem, offering scalable computing resources, extensive data storage, and networking capabilities. Providers like Amazon Web Services (AWS), Microsoft Azure, and Google Cloud Platform deliver specialized services for AI, from model training to deployment, enabling organizations to leverage cloud efficiencies for their AI initiatives.

Data Centers

The physical locales of AI computational activities, data centers, are equipped with the necessary servers and IT infrastructure to support cloud-based AI operations, emphasizing the importance of scalability and access in the AI landscape.

Training and Inference Phases

The Infrastructure Layer underpins both the training phase, where AI models learn from data, and the inference phase, where these trained models are applied to generate predictions or insights, highlighting its role in the complete AI model lifecycle.

In essence, the Infrastructure Layer is the technological cornerstone of Generative AI, providing a robust and efficient framework necessary for the development, deployment, and scaling of AI models. Its elements work in concert to ensure that AI applications can operate effectively across various domains, underscoring the critical importance of this layer in the broader AI ecosystem.

Exploring the Horizons of Generative AI Deployment

Generative AI Deployment, particularly involving foundation models for vision, speech, and language, plays a crucial role in accelerating AI development. These models serve as a springboard for a range of AI applications, facilitating rapid advancement and customization in various industries. Let's discuss the deployment options of Model Training, Fine-Tuning, Retrieval-Augmented Generation (RAG), and Prompt Engineering, and how they contribute to the trend towards more customized AI models.

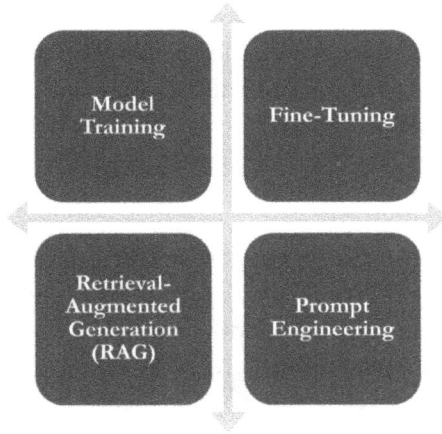

Figure 13-2: The Four Deployment Options of Generative AI

1. **Model Training**: This is the foundational step in AI deployment where a model learns from data. In the case of foundation models for vision, speech, and language, this involves processing massive datasets to understand patterns and relationships. For example, a vision model might be trained on millions of images to recognize and interpret visual data, while a language model might learn from extensive textual data. The training phase is crucial for developing a model's base capabilities before it can be fine-tuned for specific tasks.

2. **Fine-Tuning**: Once a foundation model is trained, fine-tuning tailors it to specific applications or industries. This process involves additional training on a more targeted dataset. For instance, a language model could be fine-tuned on legal documents to better serve legal industry applications, or a speech recognition model could be adapted to understand medical terminology for healthcare uses. Fine-tuning enhances a model's effectiveness and efficiency in specific contexts, aligning it more closely with industry-specific requirements.

3. **Retrieval-Augmented Generation (RAG)**: RAG combines the power of pre-trained models with external knowledge sources to generate more accurate and contextually relevant responses. This technique is particularly useful in scenarios where a model needs to pull in information from various databases or documents to respond accurately. For example, in customer service applications, RAG can help a chatbot retrieve product information from an external database to answer customer queries more effectively.

4. **Prompt Engineering**: This involves crafting effective input prompts to

guide the AI model towards desired outputs. Effective prompt engineering can significantly impact the quality of the generated content. In language models, for instance, how a question or command is phrased can determine the relevance and accuracy of the model's response. This skill is becoming increasingly important as a way to harness the full potential of powerful, but sometimes unpredictable, generative AI models.

The move towards customized AI models marks a significant trend in the deployment of generative AI, signaling a departure from generic solutions towards models that are intricately adapted to specific industry needs. This evolution not only enhances the relevance and utility of AI across various sectors but also improves efficiency by focusing model capabilities on pertinent data and tasks. The result is a suite of tailored solutions that promise superior performance and heightened user satisfaction, underscoring the transformative potential of customized AI models in driving forward industry-specific innovations.

Model Training: The Foundation of GenAI

Model Training in AI development is akin to nurturing the foundational stage of a living organism, where a model is meticulously built from its initial stages, much like growing a plant from a seed. This crucial process establishes the AI's core capabilities and intelligence. It is during this phase that the AI model learns and internalizes the basic patterns and knowledge it will use to interpret and interact with the world, forming the bedrock upon which more complex AI functions are developed.

How it Works

- **Data Collection**: The first step is gathering a diverse and large dataset. The richness and variety in data shape the AI's understanding, much like diverse experiences influencing a young mind.

- **Algorithm Selection**: This involves choosing the right learning approach, such as supervised, unsupervised, or reinforcement learning. The selection impacts how the model will process information and learn from data.

- **Training Process**: Here, the AI model learns to recognize patterns and make decisions based on the data. This iterative computational process is where the model's accuracy and efficiency are honed.

When to Use Model Training

- **New Domains**: Essential when exploring new areas where pre-existing models don't apply, such as pioneering medical diagnoses.

- **Unique Data Sets**: When dealing with data specific to certain requirements, like a company's customer data for predicting purchase patterns.

- **Innovation and Research**: Particularly suitable for R&D in AI, testing new theories or models.

Advantages

- **Customization**: Allows for high degrees of customization specific to the task.

- **Control**: Provides complete control over the entire learning process.

- **Potential for Breakthroughs**: Can lead to groundbreaking models that redefine AI in specific fields.

Challenges

- **Resource-Intensive**: Demands significant computational power and time.

- **Data Dependency**: Relies heavily on the quality and volume of data.

- **Risk of Failure**: There's a higher chance of suboptimal performance, particularly in new and unexplored domains.

Real-World Examples

- **Geographic-Specific Weather Prediction**: Developing an AI model for predicting weather patterns in a unique geographic area would require a new model trained on specific climatic data.

- **Large Language Models**: The development of models like OpenAI's GPT-3 involved extensive training on vast internet text datasets, enabling them to understand and generate human-like text.

Model Training is a vital step in AI development, offering customization and potential for innovation. However, it requires substantial resources and carries inherent risks, making it a suitable approach for situations demanding bespoke solutions or where AI is being pushed into new frontiers.

Fine-Tuning: The Art of Specialization

Fine-Tuning in AI is akin to refining a skilled artist's proficiency in a particular style. It's the process of adapting a general-purpose, pre-trained AI model to specialize in a specific task or dataset. This crucial step tailors the model to meet unique, specialized requirements.

How it Works

- **Starting with Pre-Trained Models**: The process initiates with a model that has been trained on a broad dataset, possessing a general understanding but not fine-tuned for specialized tasks.

- **Specialized Training Data**: The model undergoes additional training with data specific to the desired task. It's akin to an artist learning to work with new themes and mediums.

- **Adjustments and Refinements**: Parameters of the model are fine-tuned to better grasp and execute the particular task, requiring less computational resources than the initial training phase.

When to Use Fine-Tuning

- **Task-Specific Applications**: Perfect for aligning the general model with specific requirements, such as adapting a language model for medical terminology.

- **Limited Resources**: Ideal when extensive resources for full model training aren't available.

- **Enhancing Model Performance**: Suitable for boosting a pre-trained model's accuracy in specific areas.

Advantages

- **Efficiency**: Requires fewer resources than training a model from scratch.

- **Quick Results**: Leads to improved performance in less time due to the model's foundational knowledge.

- **Targeted Performance**: Focuses on enhancing the model's abilities in particular areas.

Challenges

- **Dependence on Base Model**: The success of fine-tuning hinges on the quality of the pre-trained model.

- **Overfitting Risks**: Overfitting can occur when fine-tuning on very specific or small datasets.

- **Limited Scope**: Improvements are bound by the base model's inherent capabilities.

Real-World Examples

- **Adapting for Language Differences**: Fine-tuning an English language sentiment analysis model for Spanish by training it with Spanish data is more efficient than building a new model.

- **Specializing AI for Culinary Conversations**: Refining OpenAI's GPT model for a cooking chatbot with culinary datasets. This transforms the model into a specialized assistant capable of detailed culinary discussions.

Fine-tuning in AI is a process of specialization, transforming a general-purpose model into one with specific expertise. Balancing efficiency and performance enhancement, it's a strategic approach for targeted improvements where the foundational model is solid, but additional specialized skills are needed.

Retrieval-Augmented Generation (RAG): Broadening Perspectives

Retrieval-Augmented Generation (RAG) marks a significant leap in generative AI by enhancing traditional large language models (LLMs) with the ability to integrate external knowledge sources. It's akin to empowering an AI 'scholar' with instant access to a vast, ever-evolving library of information, extending far beyond its original training data.

How RAG Works

- **Integration with External Databases**: RAG models blend pre-trained language models with dynamic data retrieval from external sources, akin to consulting a continuously updated database.

- **Querying and Fetching Relevant Information**: Upon receiving a query, RAG systems sift through external sources to fetch pertinent information, crucial for providing current and accurate responses.

- **Synthesizing Retrieved Data**: The model combines this newly retrieved data with its existing knowledge, crafting comprehensive and informed responses.

When to Use RAG

- **Complex Question Answering**: Suited for queries involving current events or specialized knowledge that extends beyond the model's training data.

- **Dynamic Information Requirements**: Ideal for fields where information is constantly updated, such as news, finance, or medical research.

- **Enhancing Model Capabilities**: Useful for expanding the scope of pre-trained models, particularly in delivering context-rich responses.

- **Mitigating Hallucinations**: In scenarios where accuracy is paramount, RAG helps minimize false information generation by sourcing from reliable external data.

Advantages

- **Extensive Information Access**: Provides in-depth and relevant answers by tapping into a broader information base.

- **Current and Accurate Responses**: Continuously updates its knowledge, ensuring responses are timely and factual.

- **Versatility**: Adaptable to various domains needing a mix of depth and breadth in information processing.

Challenges

- **Reliance on External Sources**: The model's efficacy is tied to the quality and accessibility of external databases.

- **Complex Integration**: Merging retrieval systems with AI models is technically challenging and resource-heavy.

- **Balancing Relevance with Accuracy**: It can be challenging to ensure that retrieved information is both pertinent and accurate, especially in rapidly changing knowledge areas.

Real-World Examples

- **Medical Diagnosis Assistant**: For an AI system in healthcare requiring access to the latest research and patient data, RAG enables integration of current medical information from various databases and journals.

- **AI-Powered Academic Research Assistants**: These assistants utilize RAG to access academic databases, providing rapid, relevant information for researchers. They do more than retrieve; they synthesize and summarize complex data, aiding significantly in literature reviews and suggesting new research directions.

RAG in AI represents a crucial development, significantly enhancing the capabilities of language models. By leveraging external databases, RAG models deliver detailed, up-to-date, and context-rich responses, proving invaluable in areas where knowledge is extensive and continually evolving. Yet, their success depends on the quality of external sources and the intricacy of system integrations, presenting unique challenges.

Prompt Engineering: The Key to Unlocking Potential

Prompt Engineering, often an underappreciated aspect in the generative AI field, is a subtle yet powerful technique for extracting remarkable capabilities from pre-trained models. Its power lies not in altering AI's internal mechanics, but in skillfully guiding its output through well-crafted prompts.

Prompt Engineering is akin to a maestro directing an orchestra; the quality of the output heavily depends on the conductor's skill. In this context, AI is the orchestra, and prompts are the conductor's cues. A well-designed prompt can steer AI to generate outputs that might seem impossible at first glance.

The Underestimated Power of Prompt Engineering

- **Perceived Simplicity:** The underestimation of Prompt Engineering often stems from its apparent simplicity. On the surface, it seems as straightforward as typing a query into a search engine — a task perceived as requiring little skill or thought. This perception, however, masks the intricate artistry and deep understanding needed to craft a prompt that precisely guides AI towards generating desired response. The skill lies not in the act of typing but in the subtlety of language used, the creativity required to craft effective prompts, the understanding of the AI's processing, and the ability to predict how different prompts will shape the output. This complexity is hidden behind the seemingly simple act of writing a prompt, leading many to undervalue the expertise required in this field.

- **Lack of Engineering Rigor:** Another key reason for the underestimation is the historical approach to Prompt Engineering. Unlike traditional engineering disciplines, which are characterized by structured methodologies and rigorous training, Prompt Engineering has often been approached more as an art than a science. This lack of formal structure and the perception of it as a more intuitive and less technical discipline contribute to its undervaluation. In many instances, creation of prompts has been more about trial and error and less about applying systematic, principled approaches. This absence of recognized standards and methodologies in Prompt Engineering has led to a perception that it lacks complexity and depth typically associated with other engineering fields.

The distinction between basic prompting and expert Prompt Engineering is similar to the difference between a casual conversation and a persuasive speech. While most people can engage in basic dialogue, crafting a speech that moves and influences an audience requires a deeper understanding of language, psychology, and rhetoric.

Filling Gaps of Prompt Engineering

Effective Prompt Engineering is both an art and a science. It involves an understanding of the AI model's capabilities and limitations, the nuances of language, and the ability to anticipate how a model will interpret and respond to different prompts. This skill set is not inherent; it requires practice, experimentation, and a keen understanding of AI behavior.

To address this gap and elevate the practice of Prompt Engineering, resources like the groundbreaking book "Prompt Design Patterns: Mastering the Art and Science of Prompt Engineering" are invaluable. This book offers a structured and systematic approach to Prompt Engineering, much like how design patterns in software engineering provide a framework for building high quality software.

When to Use Prompt Engineering: Prioritizing Efficiency and Mastery

- **The First Line of Approach:** Prompt Engineering should be considered the first line of approach in the toolkit of AI optimization techniques. Before delving into the more resource-intensive methods like model training or fine-tuning, or the more complex RAG, exploring the potential of Prompt Engineering is advisable. In many cases, the artful and strategic crafting of prompts can effectively address your needs without additional investment required for other methods.

- **Mastery Is Key:** The effectiveness of Prompt Engineering hinges on mastering its nuances — understanding both the art of language and the science of AI behavior. This mastery allows one to navigate vast capabilities of a pre-trained model and direct it towards desired outcomes with precision. By refining this skill, you can often achieve your objectives with Prompt Engineering alone, negating the need for more costly and time-consuming approaches.

- **Cost-Effective:** Prompt Engineering stands out as the most economical option among AI optimization strategies. It bypasses the need for extensive datasets, additional computational resources, and the time required for training or fine-tuning models. In a scenario where budget and resources are constraints, Prompt Engineering not only offers a viable solution but often the most efficient one.

Scenarios Ideal for Prompt Engineering

- **Creative and Dynamic Output Generation**: Whether it's generating unique content, creative writing, or dynamic responses, Prompt Engineering allows for a high degree of creativity and specificity.

- **Quick Solution Testing**: When speed is of the essence, and you need to test various approaches or get immediate results, Prompt Engineering provides a rapid way to iterate and find solutions.

- **Limited Resource Environments**: In situations where additional resources for training or fine-tuning are unavailable, Prompt Engineering becomes not just the first option but potentially the only viable one.

Emphasizing the Most Economical, Often Most Effective Route

It's important to highlight that while Prompt Engineering is the most cost-effective method, it's often also the most effective. The ability to harness the full capabilities of a sophisticated AI model through carefully designed prompts can yield surprisingly powerful results. This approach, however, requires an understanding that crafting effective prompts is a skill — one that involves both creative and analytical thinking.

Prompt Engineering should be the starting point in any AI optimization endeavor. It offers a unique blend of cost-effectiveness and potency, especially when mastered. For many AI applications, the solution lies not in building or retraining models but in the smart use of existing ones through the art and science of Prompt Engineering.

Advantages

- **Efficiency**: Does not require additional training or computational resources, making it highly efficient.

- **Flexibility**: Can be adapted to a wide range of tasks without the need to alter the underlying model.

- **Creativity**: Allows for a high degree of creative control over the model's outputs.

Challenges

- **Skill-Dependent**: The effectiveness of Prompt Engineering is heavily dependent on the user's ability to craft effective prompts.

- **Trial and Error**: Often involves a process of experimentation, which can be time-consuming. Leveraging "Prompt Design Patterns" can address this problem and save significant time.

Real-World Examples

Recently, Google introduced Gemini, a highly advanced general model, which has shown superior performance compared to OpenAI's GPT-4 in numerous academic benchmarks. Specifically, Gemini Ultra achieved a groundbreaking 90% score in the MMLU (Massive Multitask Language Understanding), surpassing human experts in various fields such as mathematics, physics, and ethics. This achievement highlights Gemini's proficiency in diverse and complex knowledge domains.

However, the evolving landscape of AI was further highlighted by Microsoft Research's innovative use of GPT-4. Through the application of new prompting techniques, derived from their Medprompt strategy originally aimed at enhancing medical query responses, GPT-4's performance in general domains significantly improved. Remarkably, these enhanced prompting techniques enabled GPT-4 to surpass even Gemini Ultra in the MMLU benchmarks, illustrating the substantial impact of expert Prompt Engineering on AI model performance. This advancement demonstrates the potential of GPT-4 and the underappreciated power of Prompt Engineering, showcasing that strategic prompting can unlock superior performance without additional model development or training.

In another instance, Anthropic's Claude 2.1, an AI model with a substantial 200K token context window, serves as a prime example of the significant enhancements Prompt Engineering can bring to AI functionality. This model underscores the vital role of strategic prompt crafting in advancing AI technology. By skillfully creating effective prompts, users are able to guide Claude 2.1 in processing information more efficiently, thus effectively navigating around its inherent limitations. This case highlights the crucial importance of Prompt Engineering in fully leveraging the potential of AI models, emphasizing that the quality of user interaction is as significant as the AI model's intrinsic capabilities.

Prompt Engineering stands as a crucial yet often overlooked component in the realm of AI. Its effectiveness lies in leveraging well-crafted prompts to tap into the latent capabilities of AI models, transforming it into a pivotal tool particularly in areas demanding creativity and innovation. As the field of AI progresses, the mastery of Prompt Engineering is set to become increasingly vital, providing a strategic avenue to garner impressive outcomes. This approach offers a distinct advantage, achieving notable results without the extensive resources typically associated with more intensive AI methodologies.

In-Depth Comparative Analysis

Aspect / Method	Model Training	Fine-Tuning	RAG	Prompt Engineering
Data Requirement	Requires large and diverse datasets.	Requires less data, often specific to the task.	Needs access to external knowledge bases or databases.	No additional data but requires well-designed prompts.
Computational Cost	High, due to the need to process large datasets.	Lower than full training, as it builds upon existing models.	Varies, can be high due to the integration of retrieval systems.	Low, as it relies on pre-trained models.
Flexibility	General-purpose, can be adapted to many tasks.	Highly specific to the task it's fine-tuned for.	Flexible, can incorporate up-to-date information from external sources.	Highly dependent on the skill in prompt design.
Time Investment	Long, as training from scratch is time-consuming.	Relatively short, as it builds on pre-existing models.	Can be lengthy due to the need for integrating and optimizing retrieval systems.	Minimal, involves creative prompt formulation.
Accuracy	Potentially high but depends on data quality and model architecture.	Can achieve high accuracy for specific tasks.	Can enhance accuracy by leveraging external information.	Dependent on the effectiveness of the prompt.
Use Cases	Broad range of applications in AI.	Task-specific applications like sentiment analysis, language translation.	Tasks requiring up-to-date information, like question answering.	Situations where customization is needed without retraining.
Scalability	Scalable but requires significant resources.	Easily scalable, especially for small-scale tasks.	Scalability depends on the efficiency of the retrieval system.	Highly scalable, limited mostly by the underlying model's capabilities.
Customizability	High, can be designed for specific tasks from the ground up.	Moderate, limited by the scope of the pre-trained model.	High, as it can pull from a variety of sources.	High, but limited by the user's ability to craft effective prompts.

Table 13-1: The Comparison of Four AI Model Learning Methods

Efficiency and Flexibility: The Art of Choosing the Right Path

In the world of generative AI optimization, the choice of methodology can be likened to selecting the best route in road construction:

- **Model Training**: This is akin to building a new road. It's a process that requires significant investment in terms of resources, time, and data. While it paves the way for creating highly customized and powerful AI models, it's a substantial undertaking that's not always necessary or feasible.

- **Fine-Tuning**: This method is comparable to modifying an existing road. Here, you start with a pre-existing model (the road) and make specific adjustments to better suit your needs. It's less resource-intensive than building a new road and can be highly effective, but it's still bounded by the limitations of the original model.

- **Retrieval-Augmented Generation (RAG)**: Adding RAG to this analogy, it's like equipping the road with dynamic signposts that pull in information from various locations. RAG combines the strengths of a pre-trained model with the ability to fetch and integrate external, up-to-date information. It's more flexible than model training and fine-tuning, as it can adapt to new information. However, its efficiency depends on the integration and processing of external data sources, which can be resource intensive.

- **Prompt Engineering**: This approach is like finding a clever shortcut. It involves using smart, strategically crafted prompts to guide a pre-trained AI model to produce desired results. This method is quick, flexible, and resource-efficient, offering a way to leverage the power of advanced AI models without the need for extensive data, computational power, or time. It's an innovative way to navigate the capabilities of AI, often achieving impressive results with minimal investment.

Accuracy and Scalability: Balancing Precision and Reach

Each AI method also has its unique strengths in terms of accuracy and scalability:

- **Model Training**: When built with high-quality data, model training can achieve exceptional accuracy. However, it's a broad approach, aiming to equip AI with general capabilities that can be adapted to various tasks. The trade-off is that it may not be as finely tuned to specific tasks without additional adjustments.

- **Fine-Tuning**: This technique offers more specificity. By adjusting a pre-existing model, it can be tailored to perform exceptionally well in a particular area or task. However, the extent of its adaptability is limited by the scope of the base model.

- **Retrieval-Augmented Generation (RAG)**: RAG excels in providing up-to-date accuracy. By integrating external knowledge sources, it ensures the AI can access the latest information, making it especially useful for tasks requiring current data. However, its scalability can be impacted by the efficiency and accessibility of the external data sources it relies on.

- **Prompt Engineering**: Perhaps the most versatile of all, Prompt Engineering leverages the underlying capabilities of pre-trained models to a remarkable extent. By crafting the right prompts, one can guide AI to perform a wide range of tasks with both high accuracy and scalability. This method shines in its ability to maximize the existing power of AI models without the need for further training or extensive resources, demonstrating that sometimes, the key to unlocking AI's potential lies not in building more sophisticated models, but in interacting with them more intelligently.

Each AI method offers distinct advantages:

- **Model Training**: For new, groundbreaking applications.

- **Fine-Tuning**: When making specific improvements to existing models.

- **RAG**: For applications needing extensive, real-time information.

- **Prompt Engineering**: For efficiently leveraging existing models in creative ways.

Understanding and choosing the right method ensures you can fully harness the potential of AI, tailoring it to your specific needs and constraints.

In conclusion, while all these methods play crucial roles in the AI ecosystem, the art of Prompt Engineering, with its low cost, high efficiency, and remarkable flexibility, stands out as a highly effective yet underutilized tool. It's time for AI practitioners and enthusiasts to embrace and explore this method to its full potential, unlocking new horizons in AI applications. Remember, in the world of AI, it's not just the power of the model that counts, but also the creativity and ingenuity with which you use it. Prompt Engineering is not just a tool; it's a canvas waiting for the artist's touch.

Guiding Principles for Selecting the Right AI Tools

In the dynamic landscape of AI technology implementation, selecting the right tools is pivotal for aligning with an organization's goals and operational dynamics. To navigate this selection process effectively, certain critical criteria stand out as essential guideposts. These benchmarks ensure that AI tools not only meet technical requirements but also complement strategic objectives, operational efficiencies, and compliance mandates. Here's a closer look at these vital selection criteria:

Scalability: The Backbone of Growth

Importance: Scalability is the lifeline of any AI tool, ensuring it can adeptly manage growing workloads and expand its capacity in tandem with business evolution. An AI tool's ability to scale prevents future bottlenecks, avoiding the need for extensive reconfiguration or incurring unexpected costs as demands increase.

Considerations: Assess the AI tool's capability to scale vertically (handling larger datasets) and horizontally (supporting more parallel processes), ensuring it can adapt to future growth trajectories seamlessly.

Time-to-First Prototype: Speed as a Competitive Advantage

Importance: The agility with which a working model or prototype can be developed is crucial, particularly in fast-moving sectors. A reduced time-to-first prototype accelerates the pace of innovation, enabling swift iterations and timely product advancements.

Considerations: Prioritize AI tools that provide pre-built models, intuitive interfaces, and robust documentation, facilitating a quicker transition from concept to prototype.

Collaboration Features: Enhancing Team Synergy

Importance: In multidisciplinary environments, AI tools with strong collaboration capabilities are indispensable. They streamline workflows, enhance synergy among diverse team members, and optimize the collective development effort.

Considerations: Opt for AI tools that support seamless model, data, and insight sharing within teams and offer integration with prevalent collaboration platforms, enhancing the cohesiveness of development activities.

Compliance Standards: Safeguarding Against Regulatory Risks

Importance: Compliance with relevant standards is non-negotiable, especially in regulated industries. Ensuring AI tools adhere to specific regulatory frameworks prevents potential legal complications and safeguards the organization's reputation.

Considerations: Verify that AI tools are compliant with sector-specific regulations, such as GDPR for data protection, HIPAA for healthcare data security, or other applicable standards, ensuring legal and ethical integrity.

These criteria serve as a cornerstone for the judicious selection of AI tools, ensuring that technological adoption is not just about harnessing cutting-edge capabilities but about driving meaningful, sustainable growth. By carefully evaluating AI tools against these benchmarks, organizations can foster more efficient development processes, realize cost efficiencies, and achieve a harmonious alignment with broader business objectives.

Streamlining AI with MLOps, AIOps, and Lifecycle Management

The seamless deployment and management of AI systems are pivotal to leveraging the full spectrum of artificial intelligence in today's digital era. This encompasses a keen grasp of MLOps, AIOps, and AI Lifecycle Management—fields that are rapidly advancing and reshaping the landscape of AI application and maintenance.

MLOps: Bridging Development and Operations

MLOps stands at the confluence of machine learning, DevOps, and software engineering, heralding a new era in AI by ensuring seamless integration between development and operational deployment. This evolving discipline is pivotal for organizations aiming to leverage the full potential of machine learning technologies efficiently and sustainably.

At its core, MLOps embodies the DevOps philosophy within the realm of machine learning, focusing on enhancing the lifecycle management of ML models from conception through deployment to ongoing maintenance. It addresses critical challenges such as ensuring model reproducibility, automating ML workflows, and maintaining model performance over time.

Fundamental Components

- **Continuous Integration and Delivery (CI/CD)**: MLOps adopts CI/CD practices tailored for ML, facilitating regular updates, testing, and deployment of models to maintain consistency and efficiency.

- **Data and Model Versioning**: Essential for tracking the evolution of models and datasets, versioning supports the reproducibility and auditability of ML projects.

- **Automation**: By automating stages from data preprocessing to model deployment, MLOps minimizes manual interventions, enhancing accuracy and productivity.

- **Monitoring and Maintenance**: Ongoing surveillance of model performance helps identify and rectify issues like model drift or data anomalies, ensuring models remain effective and relevant.

Addressing Challenges

The dynamic nature of data and operational environments poses a significant challenge in keeping ML models accurate and pertinent. MLOps addresses this through robust monitoring frameworks and strategies for periodic model updates and retraining, safeguarding model integrity over time.

Tools and Technologies

MLOps is supported by a rich ecosystem of tools and platforms, including TensorFlow Extended (TFX), MLflow, and Kubeflow, which provide comprehensive capabilities for pipeline orchestration, model serving, and performance tracking.

The MLOps Lifecycle

The MLOps lifecycle encompasses a series of interconnected stages—ranging from initial problem definition and data collection to model development, testing, and deployment. This cycle is iterative, with insights from later stages feeding back into earlier phases for continual refinement and enhancement.

Future Directions

The trajectory of MLOps is marked by trends towards greater automation, deeper cloud integration, a heightened focus on ethical AI, and the development of advanced tools for model monitoring and maintenance. These directions signify a future where MLOps not only streamlines operational efficiencies but also embeds ethical considerations and adaptability at the heart of machine learning projects.

MLOps is not just a methodology; it's a transformative approach that encapsulates the lifecycle management of machine learning models, fostering collaboration, automation, and continuous improvement. As this field matures, adopting MLOps practices is becoming increasingly vital for organizations seeking to deploy scalable, efficient, and effective AI solutions.

AIOps: Revolutionizing IT Operations with AI

AIOps, or Artificial Intelligence for IT Operations, marks a transformative approach in the realm of IT management, leveraging the prowess of artificial intelligence and machine learning to streamline and enhance various operational aspects. From monitoring and analysis to incident response, AIOps aims to automate and refine IT processes, fostering an environment of proactive issue resolution and continuous improvement.

Core Functions and Purpose

AIOps stands at the forefront of automating IT operations, employing AI models to tackle common IT challenges such as anomaly detection, event correlation, and alert reduction. Its proactive issue resolution capabilities allow for the anticipation of problems before they impact services, significantly reducing downtime and enhancing service reliability.

Pillars of AIOps

Central to AIOps is data analysis, which involves sifting through logs, metrics, and performance data to detect patterns and anomalies. Machine learning algorithms play a crucial role, learning from data to enhance their predictive capabilities and suggest effective solutions over time. Automation and orchestration are also key components, enabling AIOps tools to automate responses to specific issues and manage complex workflows across multiple IT systems.

Advantages of Implementing AIOps

The adoption of AIOps brings about enhanced operational efficiency by automating routine tasks and freeing up IT personnel for more strategic initiatives. It improves the reliability of IT services through vigilant monitoring and proactive problem-solving. Additionally, the insights provided by AIOps tools aid IT leaders in making informed decisions about infrastructure and operations, based on comprehensive data analysis.

Integration with Existing Systems

AIOps is designed to integrate seamlessly with existing IT management tools and processes, augmenting and enhancing these systems without the need for replacements or extensive overhauls.

Challenges in AIOps Implementation

Implementing AIOps effectively requires addressing the challenges of data quality and volume, as high-quality, comprehensive data is essential for AI models to function accurately. The complexity of modern IT environments can also present hurdles to the seamless adoption of AIOps solutions, necessitating careful planning and strategic execution.

The Future of AIOps

The future landscape of AIOps is marked by increasing adoption as IT environments grow in complexity and the need for efficient, automated operations becomes more pronounced. Advancements in AI and machine learning technologies promise to further enhance the capabilities of AIOps solutions, making them even more essential for managing IT operations.

AIOps represents a significant leap forward in how IT operations are managed, offering a path to automate, predict, and optimize IT processes through artificial intelligence. This not only increases operational efficiency but also ensures IT services are more reliable and aligned with business objectives. As the field of AIOps continues to evolve, its integration into IT strategies is set to become indispensable for organizations navigating the complexities of the digital landscape.

AI Lifecycle Management: Ensuring Excellence from Inception to Retirement

AI Lifecycle Management represents the strategic framework for overseeing AI models throughout their entire journey, from initial concept to eventual retirement. This systematic approach is essential for developing, sustaining, and optimizing AI systems, ensuring they deliver value and maintain efficacy throughout their operational life.

Phases of AI Lifecycle Management

- **Planning and Definition**: The journey begins with setting clear objectives, defining the scope, and outlining the requirements of the AI system. Key activities include identifying the core business problem, selecting the appropriate AI methodology, and allocating necessary resources.

- **Data Collection and Preparation**: A foundational stage that involves compiling, cleansing, and structuring data. The caliber and relevance of this data are pivotal for the subsequent performance of the AI model.

- **Model Development and Training**: At this stage, AI models are crafted and refined using the prepared datasets. Selecting the right algorithms, training the models, and fine-tuning their parameters are critical steps for optimizing model performance.

- **Evaluation and Testing**: Once developed, models undergo thorough evaluation against established metrics for accuracy, efficiency, and reliability, ensuring they meet the project's goals.

- **Deployment**: Successful models are then deployed into the operational environment, which might range from integration within existing systems to hosting on cloud platforms, making them accessible for real-world application.

- **Monitoring and Maintenance**: Post-deployment, continuous oversight is crucial to ascertain that the model performs reliably over time. This phase involves routine updates, troubleshooting performance issues, and retraining the model with updated data as needed.

- **Retirement**: The lifecycle concludes with the retirement of models that no longer serve their purpose or have become obsolete, ensuring their decommissioning is handled in a responsible and ethical manner.

Challenges and Solutions

- **Data Management**: Handling the extensive datasets required for AI initiatives presents challenges. Best practices include prioritizing data quality and adhering to data privacy and security standards.

- **Model Drift and Reproducibility**: To counteract model degradation over time, continuous monitoring and periodic retraining are essential. Ensuring models can be consistently reproduced from the same datasets and algorithms underpins reliability.

- **Scalability and Performance**: AI systems must be designed to scale and perform robustly under varying loads, requiring careful architectural planning.

- **Ethical Considerations**: Throughout its lifecycle, AI management must address ethical issues, such as mitigating bias in models and maintaining transparency in decision-making processes.

Tools, Technologies, and Trends

The management of AI lifecycles is supported by an array of tools and technologies, from data analytics platforms to machine learning frameworks and MLOps solutions. Emerging trends highlight the growing integration of cloud services in AI lifecycle management and the increasing focus on ethical AI practices, reflecting the evolving landscape of AI technology and its applications.

AI Lifecycle Management is a cornerstone practice for ensuring AI projects achieve their intended outcomes, offering a structured methodology for guiding AI models from conception through to ethical decommissioning. By adhering to this comprehensive approach, organizations can maximize the impact and value of their AI investments, navigating the complexities of AI development and deployment with strategic oversight and operational excellence.

AI-Native Thinking in AI Technology Foundation

AI-native thinking in the foundation of AI technology revolves around the principle of designing and developing technology infrastructure and tools with AI as a central, integral component. This approach profoundly impacts how foundational AI technologies are conceptualized, developed, and deployed.

Core Principles of AI-Native Technology Foundation

1. **Inherent Integration of AI in Architecture**: Traditional technology infrastructures often add AI as an afterthought or a separate layer. In contrast, AI-native thinking embeds AI deeply into the technology architecture. This means AI functionalities are not merely additions but are integral to the system's core design. This integration leads to more efficient data processing, enhanced automation capabilities, and improved scalability tailored to AI's unique demands.

2. **Data Infrastructure Designed for AI**: An AI-native technology foundation prioritizes building a data infrastructure optimized for AI processes, including data collection, storage, processing, and analysis. Such infrastructures are designed to handle the volume, velocity, and variety of data essential for training and running AI models efficiently and effectively.

3. **Specialized Hardware for AI Processing**: AI-native thinking involves using or developing hardware specifically designed for AI applications, such as GPUs for parallel processing or TPUs for handling neural network computations. This specialized hardware is crucial for enhancing the performance of AI models, particularly in handling complex computations and large datasets.

4. **Scalability and Elasticity**: AI-native infrastructures are designed to be inherently scalable and elastic, capable of adjusting to varying workloads and the expanding scope of AI applications. This flexibility is essential to accommodate the growing needs of AI systems, from small-scale experiments to large-scale deployments.

5. **Seamless Integration of Emerging AI Technologies**: AI-native foundations are built with the flexibility to incorporate emerging AI technologies and methodologies effortlessly. This adaptability ensures that the foundation remains relevant and cutting-edge. Examples include the integration of advanced machine learning algorithms, quantum computing elements, or new forms of neural network architectures.

6. **Built-in Ethical and Responsible AI Mechanisms**: From the outset, AI-native foundations incorporate mechanisms to address ethical considerations and ensure responsible use of AI. This includes features for transparency, explainability, bias detection, and data privacy.

7. **Emphasis on Developer and User Experience**: AI-native technology

foundations aim to enhance the experience of both developers working with AI tools and end-users interacting with AI-powered applications. This involves intuitive interfaces, comprehensive documentation,
and user-friendly toolkits.

AI-native thinking in AI technology foundation is not just about the incorporation
of AI into existing frameworks but about reimagining technology infrastructures
where AI is a fundamental aspect. This approach facilitates the development of
more effective, efficient, and adaptable AI solutions, paving the way for advanced
and seamless AI integration into various applications and services.

Embracing the Future: AI Technology and Its Evolving Land-scape

Throughout this chapter, we have delved into the intricate landscape of AI technology, focusing on the pivotal role of generative AI and the critical elements required
for its successful deployment and management. Our exploration spanned the foundational frameworks of AI technology, the architectural intricacies of Generative
AI, and the operational disciplines of MLOps, AIOps, and AI Lifecycle Management, revealing the complex yet fascinating nature of artificial intelligence.

This journey underscores the necessity for a solid and sophisticated infrastructure
to effectively harness the potential of AI technologies. For organizations embarking
on the AI journey, embracing these principles is not merely beneficial but essential.
It ensures operational excellence, fosters innovation, and secures a competitive edge
in a landscape increasingly dominated by AI advancements.

The insights from this chapter highlight the critical importance of ongoing education, strategic flexibility, and the careful implementation of AI principles. As
the field of AI technology continues to expand and transform, the strategies and
infrastructures that underpin it must also adapt and evolve. This constant progression requires a steadfast dedication to ethical standards, operational efficiency, and
effectiveness, guaranteeing that the full transformative promise of AI is achieved.

In essence, the future of AI technology and its application across various sectors demands a proactive and informed approach. By staying abreast of developments and
integrating the core principles of AI deployment and management, organizations
can not only navigate the AI revolution but also lead it, shaping a future where AI
technologies are leveraged for the greater good, driving progress and innovation in
an ever-changing world.

Chapter 14

AI Data Foundation: The Blueprint of AI Data Readiness

"Big data is at the foundation of all the megatrends that are happening." —— Chris Lynch

I n the rapidly evolving landscape of technology, Artificial Intelligence (AI) emerges as a beacon of disruption, promising to redefine the contours of industries and the very fabric of our daily lives. As companies navigate this transformative era, they find themselves at a crossroads, grappling with a pivotal question: How can they harness the unprecedented power of AI to secure a competitive edge? The answer lies in the realm of generative AI, a revolutionary branch that is crafting new realities by generating novel, original content—from images and videos to text, code, simulations, and beyond. Yet, as we stand on the brink of this new frontier, a formidable challenge looms large: data readiness.

Data readiness, especially in the context of generative AI, is a multifaceted odyssey that ventures beyond the sheer volume of data. It's a journey that demands not just quantity but quality, not just information but insight. At the heart of generative AI's potential lies a voracious appetite for high-quality, multi-modal data, a hunger that far surpasses the needs of traditional analytics techniques. Without a robust foundation of data readiness, the promise of generative AI may remain unfulfilled, rendering powerful algorithms ineffective and their outputs unreliable.

The chapter will delve into pivotal topics essential for understanding and implementing AI data readiness, crucial for the successful adoption and operation of generative AI systems. These topics form the blueprint for organizations aiming to harness the transformative power of AI, outlining the strategic, technical, and ethical considerations necessary for leveraging AI's full potential.

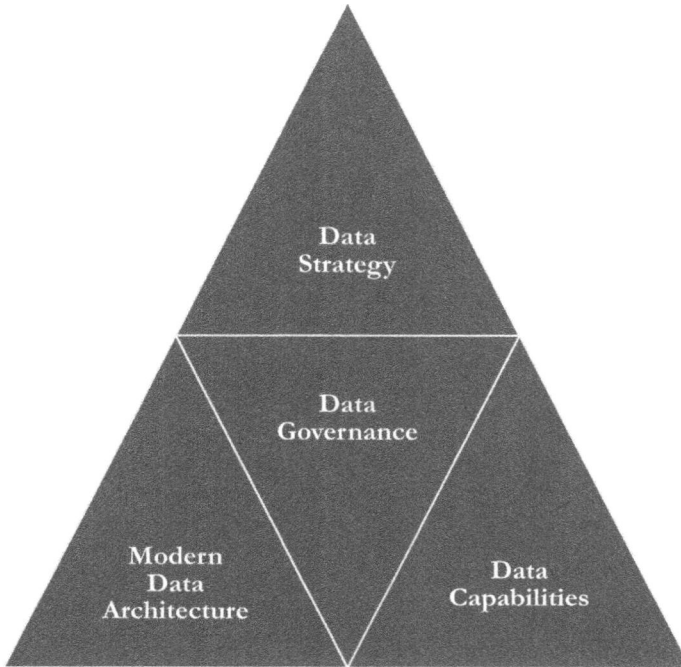

Figure 14-1: The Four Core Components of AI Data Foundation

- **Data Strategy:** We will explore the importance of a well-defined data strategy in the success of AI systems. This includes discussions on establishing clear policies for the acquisition, aggregation, and labeling of training data, ensuring data is managed in compliance with consent and privacy laws, and adhering to ethics guidelines for bias detection and mitigation. The chapter will cover efficient formats for data storage, compression, streaming, and version control to optimize model development processes, alongside secure access control policies to regulate data access rights and responsible data sharing guidelines to ensure transparency and purpose in data utilization.

- **Modern Data Architecture:** Addressing the limitations of traditional data architectures, this section will highlight the need for dynamic, scalable, and flexible data infrastructures to support the voluminous and varied data requirements of generative AI models. We will discuss how a robust and adaptable data architecture is not merely a technical requirement but a strategic necessity, facilitating the effective management and processing of multi-structured data for both the development and practical application of generative AI models in real-world scenarios.

- **Data Governance:** With the exponential increase in data needs and the distribution of datasets across hybrid infrastructures, robust data governance becomes indispensable. This segment will emphasize the role of data governance in ensuring accurate and timely inputs for generative AI models, thus enhancing output quality. We'll cover aspects of Master Data Management, Data Quality Management, Metadata Governance, and the adherence to Data Ethics and Regulations, establishing a structured and controlled data environment.

- **Data Capabilities:** Finally, the development of specialized data capabilities is essential for organizations to fully leverage the potential of generative AI. This involves a focus on cultivating advanced analytics talent, implementing self-service data access platforms, and adopting agile team constructs. Such capabilities are crucial for effectively managing, analyzing, and utilizing data to drive AI innovation and operational excellence.

Through these discussions, the chapter aims to provide readers with a comprehensive understanding of the critical elements of AI data readiness, offering a strategic blueprint for organizations to navigate the complexities of data management and governance in the age of generative AI.

Developing an AI Data Strategy

The development of a data strategy for generative AI is a critical process that must align with business objectives and the unique requirements of AI applications. In real-world scenarios, this process follows a logical flow, beginning with defining objectives and culminating in an agile, adaptable strategy.

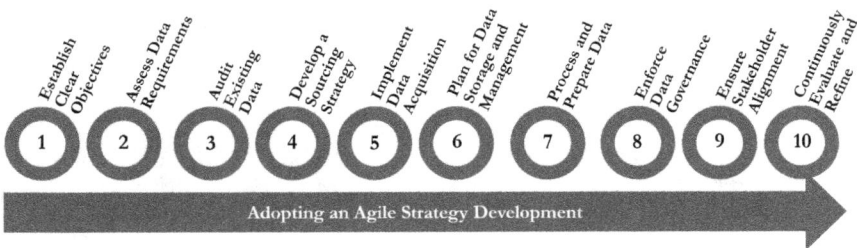

Figure 14-2: The 10-Step Agile Development of AI Data Strategy

1. Start with Clear Objectives

In the context of developing a data strategy for generative AI, starting with clear objectives is crucial. This initial step involves setting specific, measurable, and achievable goals for the AI initiatives. These objectives serve as a guiding beacon throughout the strategy development process, ensuring that every decision and action aligns with what the organization aims to achieve through its AI endeavors.

Key Elements of Setting Objectives

- **Defining AI Goals**: Clearly articulate what the organization aims to achieve with generative AI. This could range from improving customer experiences to creating new product offerings or enhancing operational efficiencies.

- **Business Alignment**: Ensure that the AI objectives are in sync with the broader business goals. This alignment guarantees that the AI initiatives contribute to the overall success and strategic direction of the organization.

- **Stakeholder Engagement**: Involve key stakeholders from various departments in setting these goals. Their insights can provide a comprehensive view of how AI can be leveraged across different areas of the business.

Guidance for Data Strategy Development

- **Decision-Making Framework**: Use the objectives as a framework for making decisions throughout the data strategy development process. This ensures consistency and focus on the end goals.

- **Prioritization**: The defined objectives help in prioritizing the data needs, technological investments, and resource allocation. It becomes easier to decide what data to collect, how to process it, and where to focus AI development efforts.

- **Success Metrics**: Establish clear metrics for success based on the objectives. These metrics could include improved customer satisfaction scores, increased revenue from new AI-driven products, or enhanced operational efficiency.

Challenges and Considerations

- **Balancing Ambition and Realism**: Objectives should be ambitious enough to drive innovation but also realistic and achievable given the organization's current capabilities and resources.

- **Dynamic Adjustments**: Be prepared to revisit and adjust these objectives as the organization progresses with its AI initiatives and as the external market and technological landscape evolve.

- **Cross-Functional Understanding**: It is vital to ensure that there is a shared understanding of these objectives across all teams involved. Misalignment can lead to fragmented efforts and diluted results.

Starting with clear objectives forms the cornerstone of an effective AI data strategy. It ensures that all subsequent efforts in data preparation, model development, and AI implementation are aligned with what the organization aims to achieve. This step lays a solid foundation for a focused and strategic approach to leveraging generative AI, driving meaningful outcomes that resonate with the organization's overall vision and goals.

2. Conducting a Targeted Assessment of Data Requirements

Once the objectives are clear, assess the data requirements specific to your intended AI applications. For example, an AI assistant generating marketing offers would require diverse data such as historical campaign assets and consumer dialogues. This process involves identifying the specific types of data necessary for the AI applications envisioned. It helps in understanding the volume, variety, and quality of data needed, ensuring that the subsequent AI models are well-informed and effective.

Key Elements of Data Requirements Assessment

- **Understanding AI Application Needs**: Analyze the intended AI applications to determine the nature of data required. For instance, if the application involves image generation, the data required would be significantly different from what is needed for natural language processing tasks.

- **Functionality-Specific Data Identification**: Each functionality of the AI application may require different types of data. For example, an AI tool designed for personalized customer interactions would need customer behavior data, transaction history, and possibly, demographic information.

- **Data Volume and Diversity**: Assess the volume (amount of data) and variety (types of data) needed. Generative AI models, particularly those based on deep learning, often require large datasets to train effectively.

Guidance for Data Strategy Development

- **Data Sourcing Plan**: Based on the assessment, develop a plan for sourcing the required data. This could involve collecting new data, purchasing data from third parties, or using publicly available datasets.

- **Gap Analysis**: Compare the identified data needs with the current data holdings to identify gaps. This analysis will guide the data acquisition and enhancement efforts.

- **Quality and Format Considerations**: Ensure that the data not only meets the quantity requirements but is also of high quality and in the right format for AI processing.

Challenges and Considerations

- **Accuracy and Relevance**: Ensure the data identified is accurate and relevant to the specific AI applications. Irrelevant or low-quality data can lead to poor AI performance.

- **Scalability and Future-Proofing**: Consider future scalability in the assessment. The data needs might evolve as the AI applications become more sophisticated or as business needs change.

- **Compliance and Ethical Concerns**: Be mindful of data privacy laws and ethical considerations, especially when dealing with sensitive or personal data.

A thorough assessment of data requirements is vital for laying a strong foundation for generative AI initiatives. It informs the organization about the kind of data needed, its sources, and the gaps that need to be filled. This step ensures that the AI models are built on a robust and relevant dataset, crucial for their effectiveness and accuracy. By carefully analyzing and understanding these requirements, organizations can make informed decisions in their journey towards adopting and benefiting from generative AI technologies.

3. Auditing Existing Data Assets

With a clear understanding of the required data, audit your existing data assets. This process involves evaluating the current data holdings of the organization in terms of volume, variety, velocity, and veracity. An effective audit provides a clear picture of the available resources and highlights the gaps between existing assets and the requirements identified in the previous step.

Key Elements of Data Asset Audit

- **Volume Assessment**: Evaluate the quantity of existing data. Generative AI models, especially those based on deep learning, often require large datasets for effective training and accurate output.

- **Variety Evaluation**: Assess the types of data available. Generative AI can leverage diverse data types like text, images, videos, and audio. A rich variety in data allows for more versatile and robust AI applications.

- **Velocity Analysis**: Consider the speed at which data is collected and updated. Real-time or frequently updated data can be crucial for applications that rely on current information, like market trend analysis or real-time personalization.

- **Veracity Check**: Ensure the accuracy and reliability of the data. This involves checking for biases, errors, and inconsistencies in the data set. High veracity is essential to build trust in the AI's outputs.

Guidance for Data Strategy Development

- **Identifying Gaps**: Use the audit results to identify gaps in the data holdings that need to be addressed to meet the AI application requirements.

- **Data Enrichment Plan**: Based on the gaps identified, develop a strategy for enriching the current data set. This could involve additional data collection, data purchasing, or partnerships for data sharing.

- **Data Quality Improvement**: If issues with data quality are identified, plan for data cleaning, standardization, and validation processes.

Challenges and Considerations

- **Complexity in Large Enterprises**: In large organizations, data might be siloed across different departments, making it challenging to get a comprehensive view.

- **Continuous Data Evolution**: Data assets are not static; they evolve over time. The audit process should be dynamic and regularly updated to reflect the most current state.

- **Compliance and Security**: Ensure that the audit respects data privacy laws and security protocols, especially when dealing with sensitive or personal data.

A thorough audit of existing data assets is a critical step in building a data strategy for generative AI. It helps in understanding the current state of data holdings, identifying gaps, and planning for necessary improvements or acquisitions. This step ensures that the foundation upon which AI models are built is robust, diverse, and aligned with the specific needs of the AI applications. By accurately assessing their current data landscape, organizations can make informed decisions that drive the effectiveness and success of their generative AI initiatives.

4. Developing a Sourcing Strategy

After identifying the gaps between current data assets and the requirements for generative AI applications, developing a sourcing strategy becomes crucial. This strategy outlines how an organization will acquire the necessary data, whether through internal means, external purchases, or other methods. A well-crafted sourcing strategy ensures that the data acquired is not only sufficient in quantity and variety but also of high quality and relevance.

Key Elements of a Data Sourcing Strategy

- **Internal Data Sourcing**: Assess the potential of internal data sources. This involves exploring various departments or systems within the organization that could provide relevant data not currently being utilized.

- **External Data Acquisition**: Identify and evaluate external data sources. This might include purchasing data from third-party providers, accessing public datasets, or establishing partnerships for data sharing.

- **Synthetic Data Generation**: Consider the creation of synthetic data, especially in cases where real data is scarce, sensitive, or expensive to acquire. Synthetic data should be realistic enough to train effective AI models.

- **Ethical and Legal Considerations**: Ensure that all data sourcing methods comply with legal standards and ethical guidelines, particularly in terms of privacy and consent.

Guidance for Data Strategy Development

- **Alignment with AI Goals**: The sourcing strategy should align with the specific goals and requirements of the AI initiatives. This alignment ensures that the data acquired will be relevant and valuable.

- **Cost-Benefit Analysis**: Evaluate the cost implications of different sourcing methods and balance them against the potential value they bring to the AI projects.

- **Quality Over Quantity**: Prioritize the quality of data over sheer volume. High-quality data leads to more reliable and effective AI models.

Challenges and Considerations

- **Data Integration Challenges**: Integrating new data sources, especially external ones, can be complex and may require significant preprocessing and transformation.

- **Scalability and Sustainability**: The sourcing strategy should be scalable and sustainable over time, considering the evolving nature of AI projects and data landscapes.

- **Data Privacy and Security**: Strictly adhere to data privacy laws and security protocols, particularly when dealing with external data sources or sensitive information.

A comprehensive data sourcing strategy is a cornerstone of a successful data strategy for generative AI. It provides a structured approach to acquiring the necessary data, ensuring it is relevant, high-quality, and obtained in a legal and ethical manner. By carefully considering both internal and external sources, and potentially synthetic data, organizations can effectively bridge the gaps in their data assets, paving the way for robust and effective AI applications. This strategy plays a pivotal role in realizing the full potential of generative AI technologies, driving innovation and competitive advantage.

5. Data Acquisition and Collection

Data acquisition and collection is the active phase of implementing the data sourcing strategy. It's where organizations gather the necessary data to fuel their generative AI models. This step is crucial because the quality and relevance of the data collected directly impact the performance and accuracy of the AI applications.

Key Elements of Data Acquisition and Collection

- **Implementing Sourcing Plans**: Execute the plans developed in the sourcing strategy, whether it involves internal data collection, external purchases, or generating synthetic data.

- **Diversified Data Collection**: Ensure the collection of a diverse set of data types (e.g., text, images, videos, etc.) to enhance the robustness and versatility of AI models.

- **Data Quality Assurance**: During collection, prioritize data quality. Ensure that the data is accurate, relevant, and free from biases that could skew AI outcomes.

- **Scalable Collection Processes**: Design data collection processes that are scalable and can adapt to increasing volumes or new types of data as AI applications evolve.

Guidance for Data Strategy Development

- **Structured Data Collection**: Develop a structured approach to data collection, ensuring consistency and reliability in the data gathered.

- **Ethical and Legal Compliance**: Adhere strictly to ethical guidelines and legal requirements, especially in handling personal or sensitive data.

- **Continuous Monitoring and Adjustment**: Regularly monitor the data collection process for gaps or inefficiencies and adjust strategies as necessary.

Challenges and Considerations

- **Integrating Diverse Data Sources**: Integration can be challenging, especially when dealing with various data formats and sources.

- **Ensuring Data Privacy and Security**: Maintain stringent privacy and security measures to protect data integrity and comply with regulatory standards.

- **Managing Data Volume**: Effectively manage the large volumes of data that generative AI systems often require, ensuring storage and processing capabilities are not overwhelmed.

Data acquisition and collection are fundamental to building a data foundation for generative AI. By focusing on quality, diversity, and compliance in data collection, organizations can ensure that their AI models are fed with the best possible data. This step requires careful planning and execution, with an emphasis on ethical practices and adaptability to changing requirements. Effective data acquisition and collection set the stage for successful AI model training and operation, ultimately driving the value and effectiveness of AI initiatives.

6. Data Storage and Management

With new data coming in, plan for efficient storage and management. Data storage and management are critical in ensuring that the data acquired for generative AI is organized, secure, and easily accessible for processing and analysis. Efficient storage and management strategies are essential for handling the large volumes of diverse data that generative AI requires. Proper data management not only facilitates smoother operation of AI models but also ensures compliance with data governance policies.

Key Elements of Data Storage and Management

- **Selecting Appropriate Storage Solutions**: Choose storage options that can handle the scale and complexity of the data. This may involve cloud-based solutions, on-premises databases, or hybrid systems.

- **Data Organization and Cataloging**: Organize and catalog data for easy retrieval and use. This includes metadata management to make the data easily searchable and usable for AI purposes.

- **Ensuring Data Security and Privacy**: Implement robust security measures to protect data from unauthorized access and breaches. Compliance with privacy laws and regulations is paramount, especially with sensitive or personal data.

- **Data Accessibility and Integration**: Ensure that the data is accessible to relevant stakeholders and systems. This involves integrating data from various sources and formats into a unified system that AI tools can readily utilize.

Guidance for Data Strategy Development

- **Scalability and Flexibility**: Develop storage and management systems that are scalable to accommodate growing data needs and flexible enough to handle different types of data.

- **Backup and Recovery Plans**: Implement backup strategies and disaster recovery plans to prevent data loss and ensure business continuity.

- **Data Lifecycle Management**: Manage the entire lifecycle of the data, from acquisition and storage to usage and eventual archiving or deletion.

Challenges and Considerations

- **Handling Large Data Volumes**: Managing the large volumes of data required for generative AI can be challenging, requiring robust infrastructure and efficient data management practices.

- **Data Integration Complexity**: Integrating data from various sources and formats can be complex, necessitating advanced tools and techniques.

- **Cost Management**: Storage and management solutions can be costly, especially for large volumes of data. Balancing cost with efficiency and security is crucial.

Effective data storage and management are foundational to the success of generative AI initiatives. By carefully selecting appropriate storage solutions, organizing and securing data, and ensuring its accessibility and integration, organizations can create an efficient data ecosystem. This ecosystem not only supports the operational needs of AI models but also aligns with data governance and compliance requirements. Navigating the challenges of data volume, integration, and cost is key to building a robust data management framework that underpins successful AI applications.

7. Data Processing and Preparation

Data processing and preparation are essential stages in making data suitable for generative AI applications. This step involves transforming raw data into a format and quality that AI models can effectively use for training and analysis. Proper data processing and preparation enhance the model's performance, accuracy, and reliability.

Key Elements of Data Processing and Preparation

- **Data Cleaning**: Remove inaccuracies, inconsistencies, and duplicates from the data. This step is crucial to ensure that AI models are not trained on faulty data, which can lead to inaccurate outputs.

- **Data Transformation**: Convert data into formats suitable for AI processing. This might involve normalization, aggregation, and encoding of data, especially when dealing with diverse data types.

- **Feature Engineering**: Identify and create relevant features that improve the model's ability to learn from the data. This step involves understanding the underlying patterns and characteristics in the data that are significant for the AI models.

- **Data Labeling and Annotation**: For supervised learning models, label the data accurately. Data labeling is critical in training models to recognize patterns and make predictions.

Guidance for Data Strategy Development

- **Automated Processing Tools**: Implement automated tools and technologies for efficient and consistent data processing. This is especially important given the large volumes of data used in generative AI.

- **Quality Assurance**: Establish quality checks and balances to ensure data integrity throughout the processing phase.

- **Scalability of Processes**: Design data processing and preparation workflows that can scale with the increasing volumes and complexity of data.

Challenges and Considerations

- **Balancing Quality and Quantity**: Processing large volumes of data without compromising quality is a significant challenge. Ensuring that data is both high in quantity and quality is vital for effective AI models.

- **Handling Complex Data Types**: Diverse data types (like images, text, and audio) require specialized processing techniques, which can add complexity to the preparation phase.

- **Time and Resource Intensive**: Data processing and preparation can be time-consuming and resource-intensive, necessitating adequate planning and allocation of resources.

Data processing and preparation are crucial for transforming raw data into a valuable asset for generative AI. By cleaning, transforming, and appropriately preparing data, organizations can significantly enhance the effectiveness of their AI models. This step requires a careful balance between quality and quantity, the use of automated tools for efficiency, and scalable processes to handle growing data needs. Properly processed and prepared data is the foundation upon which reliable and accurate AI models are built, making this step a cornerstone in the development of a robust AI data strategy.

8. Data Governance

Data governance in the context of generative AI involves establishing policies, procedures, and standards to manage and use data responsibly and effectively. Effective data governance ensures data quality, security, privacy, and compliance with regulatory requirements. It is essential for maintaining trust in AI systems, particularly when dealing with sensitive or personal data.

Key Elements of Data Governance

- **Policy Development and Implementation**: Develop comprehensive policies that define how data is to be handled, accessed, and used within the organization. These policies should cover data security, privacy, ethical use, and compliance with laws like GDPR or HIPAA.

- **Data Quality Management**: Establish systems and protocols to ensure and maintain the quality of data. This includes setting standards for data accuracy, completeness, consistency, and reliability.

- **Access Control and Security**: Implement robust access control measures to ensure that only authorized personnel can access sensitive data. Security protocols should protect data from unauthorized access, breaches, and other cyber threats.

- **Compliance and Legal Requirements**: Ensure that data governance policies comply with all relevant legal and regulatory requirements. Stay updated with changes in data protection laws and adjust governance policies accordingly.

- **Ethical Data Usage**: Develop guidelines for ethical data usage, especially when dealing with data that could have privacy implications or could be used in ways that raise ethical concerns.

Guidance for Data Strategy Development

- **Integrate Governance with AI Strategy**: Data governance should be an integral part of the overall AI strategy, ensuring that data practices align with organizational goals and ethical standards.

- **Stakeholder Involvement**: Involve stakeholders from various departments in developing and implementing data governance policies. This inclusive approach ensures broad acceptance and adherence to the governance framework.

- **Continuous Monitoring and Auditing**: Regularly monitor and audit data practices to ensure compliance with governance policies. This also helps in identifying and addressing any governance-related issues promptly.

Challenges and Considerations

- **Balancing Security with Accessibility**: Finding the right balance between securing data and making it accessible for AI development can be challenging. Overly restrictive access controls can hinder AI innovation, while lax security can pose risks.

- **Evolving Regulatory Landscape**: The legal landscape around data privacy and protection is continuously evolving. Keeping up with these changes and ensuring compliance can be complex and resource-intensive.

- **Managing Data Bias and Ethics**: Ensuring that data governance address-es issues of bias and ethical use of AI is critical, especially given the broad impact AI can have on society.

Data governance is a critical pillar in the development of a data strategy for gen-erative AI. It provides the framework within which data is securely, ethically, and legally managed. Effective governance not only ensures compliance and maintains data quality but also builds trust in AI systems among users and stakeholders. By embedding robust data governance practices into their AI strategy, organizations can navigate the complexities of data management and harness the full potential of their AI initiatives.

9. Ensuring Stakeholder Alignment and Co-creation

Ensuring stakeholder alignment and co-creation is pivotal in the development of a data strategy for generative AI. This process involves engaging various stakeholders across the organization to collaboratively contribute to and support the AI data strategy. Stakeholder alignment guarantees that the strategy is comprehensive, re-alistic, and attuned to the diverse needs and perspectives within the organization.

Key Elements of Stakeholder Alignment and Co-creation

- **Identifying and Engaging Stakeholders**: Identify key stakeholders from different departments such as IT, business units, legal, compliance, and op-erations. Engage them early in the process to gather insights, expectations, and requirements.

- **Cross-Functional Collaboration**: Foster collaboration between different departments and teams. This ensures that the data strategy benefits from a range of expertise and viewpoints, leading to a more robust and effective outcome.

- **Communication and Transparency**: Maintain open communication channels throughout the strategy development process. Transparency about goals, processes, and decisions helps in building trust and buy-in from all stakeholders.

- **Joint Problem-Solving and Decision Making**: Involve stakeholders in problem-solving and decision-making processes. This co-creative approach ensures that the strategy addresses real business needs and challenges.

Guidance for Data Strategy Development

- **Aligning Strategy with Business Objectives**: Ensure that the data strategy aligns with the overall business objectives and goals of the organization. Stakeholder input is crucial in achieving this alignment.

- **Understanding Operational Realities**: Gain insights into the operational realities and constraints from stakeholders. This understanding helps in developing a practical and implementable data strategy.

- **Prioritizing Resources and Efforts**: Use stakeholder input to prioritize initiatives, allocate resources effectively, and sequence activities in the data strategy.

Challenges and Considerations

- **Balancing Diverse Interests**: Managing and balancing the diverse interests and viewpoints of various stakeholders can be challenging. Finding common ground and consensus is key.

- **Change Management**: Implementing a new data strategy can involve significant changes in processes and operations. Effective change management is crucial to ensure stakeholder buy-in and smooth transition.

- **Continuous Engagement**: Maintaining ongoing stakeholder engagement throughout the development and implementation of the data strategy is essential but can be resource-intensive.

Ensuring stakeholder alignment and co-creation is essential for developing a successful data strategy for generative AI. It facilitates a comprehensive, realistic, and effective strategy that is well-supported across the organization. Through cross-functional collaboration, open communication, and joint decision-making, stakeholders collectively contribute to a data strategy that not only aligns with business objectives but also addresses operational realities and challenges. This collaborative approach is critical in ensuring the successful adoption and implementation of the AI data strategy.

10. Continuous Data Evaluation and Refinement

Continuous data evaluation and refinement are essential components of a dynamic data strategy, especially in the context of generative AI. This ongoing process ensures that the data strategy remains effective, relevant, and aligned with the evolving needs of AI applications and business objectives. In the fast-paced realm of AI and data science, what is effective today may not suffice tomorrow, making continuous evaluation and adaptation a necessity.

Key Elements of Continuous Data Evaluation and Refinement

- **Regular Data Quality Assessments**: Conduct routine assessments to ensure the data maintains high quality standards. This includes checking for accuracy, completeness, and relevance.

- **Monitoring Data Relevance**: Regularly review the data to ensure it remains relevant to the current AI applications. As AI models evolve, so too might their data requirements.

- **Adapting to New Data Sources and Types**: Stay open to incorporating new data sources and types. The emergence of new data or changes in existing data sources can offer opportunities for enhancing AI applications.

- **Feedback Loops**: Establish feedback loops with data users, including data scientists, AI model developers, and business stakeholders, to gather insights on data usability, challenges, and potential improvements.

Guidance for Data Strategy Development

- **Data Strategy as a Living Document**: Treat the data strategy as a living document that evolves over time. Regular updates should reflect changes in business strategy, technological advancements, and market trends.

- **Agility in Data Operations**: Maintain agility in data operations to quickly respond to new findings or changes in the data landscape.

- **Balancing Stability and Change**: While it's important to adapt and evolve, there's also a need to maintain a certain level of stability in data practices to ensure continuity and reliability.

Challenges and Considerations

- **Resource Allocation for Ongoing Evaluation**: Continuous evaluation requires dedicated resources, which might be challenging to allocate in resource-constrained environments.

- **Managing Change**: Frequent changes in data strategy can disrupt ongoing operations. Managing these changes effectively requires careful planning and communication.

- **Data Lifecycle Management**: As data evolves, managing its lifecycle – from acquisition to archiving or deletion – becomes increasingly complex.

Continuous data evaluation and refinement are crucial for maintaining an effective data strategy in the dynamic field of generative AI. This ongoing process ensures that the data strategy stays aligned with the latest AI developments, business goals, and data landscapes. By regularly assessing data quality, relevance, and incorporating feedback, organizations can ensure their data strategy remains robust, agile, and fit for purpose. This continual process of evaluation and adaptation is key to leveraging the ever-evolving capabilities of AI and extracting maximum value from data assets.

Adopting an Agile and Adaptable Strategy Development

In the rapidly evolving landscape of artificial intelligence, adopting an agile and adaptable approach to data strategy development for generative AI is not just beneficial but essential. This dynamic environment, characterized by swift technological advancements and shifting market demands, necessitates a strategy that is both flexible and resilient. An agile data strategy empowers organizations to stay ahead of the curve, enabling them to swiftly capitalize on new opportunities and navigate emerging challenges with ease.

The foundation of an agile strategy is its iterative development process. This approach allows for continuous refinement and adaptation based on a cycle of feedback, new insights, and evolving requirements. By staying informed of the latest developments in AI and data management technologies, organizations can ensure their strategies are responsive and capable of integrating new tools and methodologies as soon as they become available. Establishing robust feedback mechanisms is vital, drawing on insights from a diverse array of stakeholders, including data scientists, IT personnel, end-users, and business leaders, to guide strategic adjustments.

Key to the agile approach is the integration of risk management and maintaining the flexibility to pivot or modify the strategy in response to both anticipated and unforeseen challenges. Balancing short-term goals with a long-term vision is crucial; immediate objectives should support the broader aims of AI initiatives, ensuring alignment with the organization's overall direction and goals. Adopting a modular approach, where the strategy is developed in adjustable components, allows for quick updates and adjustments without necessitating a complete overhaul, thereby maintaining momentum and focus.

However, transitioning to an agile and adaptable strategy is not without its challenges. Organizations must navigate the uncertainty inherent in rapid technological change, making decisions with incomplete information and balancing the need for innovation with caution. Efficiently allocating resources in such a fluid environment can be complex, as priorities and objectives may shift rapidly. Moreover, fostering an agile mindset across traditionally structured organizations requires a cultural shift that can be challenging to implement and sustain.

In synthesizing a robust data strategy for generative AI, several key elements emerge as critical:

- **Holistic Approach:** The strategy must encompass all aspects of data management, including acquisition, storage, processing, governance, and continuous improvement, ensuring a comprehensive and effective framework that supports AI applications.

- **Alignment with Business Goals:** Every aspect of the data strategy should align with and support the organization's overarching business objectives, ensuring that AI initiatives contribute meaningfully to the organization's success.

- **Adaptability and Flexibility:** The fast-paced evolution of AI and data science demands that strategies be adaptable and flexible, capable of responding to new challenges and opportunities as they arise.

- **Stakeholder Engagement:** Collaboration and active engagement with stakeholders across various departments are essential for crafting a strategy that is both realistic and broadly supported, enhancing its effectiveness and practicality.

Looking forward, organizations must position themselves to embrace technological advancements continually, adapting their strategies to incorporate new tools, techniques, and data types that enhance AI applications. The AI landscape is in constant flux, with new models, applications, and approaches emerging regularly. A proactive, informed stance allows organizations to quickly adapt their strategies, ensuring they remain relevant and effective.

In conclusion, developing a data strategy for generative AI is a comprehensive, multifaceted endeavor that demands attention to detail, foresight, and a commitment to ethical standards. It is a dynamic and continuous process that recognizes data as a pivotal asset for driving AI innovation and achieving a competitive advantage. By thoughtfully planning and regularly updating their data strategy, organizations can harness the transformative power of generative AI, transforming data into actionable insights, innovation, and sustained growth in the digital age.

A Modern Data Architecture Powers Generative AI

In today's rapidly advancing technological landscape, generative AI (Gen AI) represents a paradigm shift in data processing and utilization. Unlike traditional business intelligence systems or classic machine learning approaches, Gen AI transcends mere data analysis or pattern recognition. It embarks on creating novel, original synthetic data, encompassing a broad spectrum of outputs like images, text, code, video, and more. This innovative capability of Gen AI heralds a new era in digital creativity and problem-solving, enabling machines to generate content that was once the sole domain of human creativity.

However, this revolutionary technology brings forth unique challenges and demands in terms of data architecture. The traditional data architectures, primarily designed for storing, retrieving, and processing static data sets, are no longer sufficient. Gen AI requires dynamic, scalable, and flexible data infrastructures capable of handling the enormous volume and variety of data involved in training and deploying these advanced models.

This need for a robust and adaptable data architecture is not merely a technical requirement but a strategic imperative for businesses and organizations looking to harness the full potential of Gen AI. The ability to effectively manage and process vast amounts of multi-structured data not only fuels the development and refinement of Gen AI models but also enables the deployment of these models in real-world applications. From personalized product configurations to automated content creation, the applications of Gen AI are vast and varied, making it a key driver of innovation and competitive advantage in numerous industries.

Therefore, understanding and implementing a modern data architecture is crucial for any entity venturing into the realm of Gen AI. Such an architecture should not only support the intensive data demands of Gen AI but also align with the organization's overall data strategy and business objectives. In this context, exploring the components, processes, and best practices of a modern data architecture becomes an essential step towards unlocking the transformative power of generative AI.

Five Data Architecture Patterns

The evolving landscape of generative AI (GenAI) necessitates a robust and flexible data architecture. A Cloud-native hybrid data environment emerges as a key solution to address the substantial data volumes and complex processing demands inherent in GenAI applications, offering both cost-effectiveness and scalability. Here, we explore five leading data architecture patterns that are instrumental in supporting the diverse needs of GenAI:

Pattern	Key Features	Advantages	Typical Use Cases
Data Lakes on Cloud Object Storage	▪ Centralized repository for multi-structured raw data ▪ Scalable, cost-effective storage ▪ Seamless data science tool integration	▪ Flexible data handling ▪ Scalability and cost efficiency ▪ Facilitates rapid processing and analysis	▪ Diverse data type model development ▪ Big data analytics ▪ Data exploration
Cloud Data Warehouses	▪ Structured data organization ▪ Optimized storage with columnar storage ▪ Advanced query acceleration	▪ Enhanced analytics and reporting performance ▪ Efficient storage management ▪ Complex query support	▪ Business intelligence reporting ▪ Structured data analysis ▪ Machine learning with structured datasets
Feature Stores	▪ Specialized ML feature hubs ▪ Centralized storage, management, and serving ▪ Features accessible via APIs and SQL	▪ Accelerates model development ▪ Prevents data processing fragmentation ▪ Promotes feature reuse	▪ Machine learning model development ▪ Feature engineering ▪ Rapid ML model prototyping and deployment
Data Mesh	▪ Decentralized data management ▪ Domain-specific team management of data sets ▪ Complements centralized governance	▪ Innovation and autonomy fostered ▪ Domain-specific control ▪ Scalable data governance	▪ IoT data stream management ▪ Domain-specific data applications ▪ Decentralized analytics and reporting
Data Fabric	▪ Integrated data and connectivity layer ▪ Streamlined data access across environments ▪ AI and ML for metadata management	▪ Unified data access and view ▪ Real-time cross-source data access ▪ Enhanced data discovery and governance	▪ Cross-domain data integration ▪ Real-time distributed analytics ▪ Comprehensive data management in large orgs

Table 14-1: The Comparison of Five Data Architecture Patterns

- **Data Lakes on Cloud Object Storage**: Data lakes provide a centralized repository for storing vast volumes of multi-structured raw data, such as images, texts, and sensor streams. Utilizing cloud object storage for these data lakes offers scalable and cost-effective storage solutions. They are integral for GenAI, facilitating seamless integration with data science tools, essential for processing and model development.

- **Cloud Data Warehouses**: These are pivotal for the structured organization of key datasets, optimizing storage through columnar storage techniques and advanced query acceleration. Cloud data warehouses significantly enhance performance for business reporting, analytics, and the development of machine learning models.

- **Feature Stores**: Functioning as specialized hubs for machine learning (ML) features, feature stores centralize the storage, management, and serving of ML features. They address the issue of fragmented data processing efforts by making reusable features readily accessible through APIs and SQL, thereby accelerating the pace of model development.

- **Data Mesh**: This decentralized data management strategy enables domain-specific teams to independently manage the quality, access, and consumption of high-value data sets, such as IoT data streams. Data Mesh works alongside centralized data governance models to promote innovation and autonomy within specific domains.

- **Data Fabric**: Adding to this environment is the concept of a Data Fabric, which provides a cohesive and integrated layer of data and connectivity across a diverse range of data environments. Data Fabric is designed to streamline data access and sharing in a distributed data landscape, offering a unified view and access to data regardless of its location or format. It is particularly relevant for Gen AI applications that require real-time access to data across various sources and formats. Data Fabric leverages advanced technologies like AI and machine learning for metadata management, ensuring efficient data discovery, quality, and governance across the entire data ecosystem.

This hybrid Cloud-native data environment combines the flexibility needed for handling unstructured data with the structured approach necessary for analytics and governance. This duality is essential for fueling a wide array of Gen AI use cases, from personalized product configurations to automated report generation, and beyond. The environment's inherent scalability and adaptability make it a cornerstone for organizations looking to leverage the power of GenAI while maintaining control over their data infrastructure and costs.

Data Governance is the Foundation

With data needs exploding and datasets distributed across hybrid infrastructure and business domains, sound data governance ensures generative AI models access accurate, timely inputs that fuel quality outputs. Without governance guardrails, data strategies risk losing direction and data architecture descending into chaotic complexity.

Master Data Foundation

Well governed master data provides a reliable base for generative AI models to build on. Master data encompasses business critical entities like customers, products, organizational hierarchies etc. Golden records in master data hubs ensure a single source of truth for attributes that contextualize generated content like names, locations, prices etc. Statement balances in customer master help personalize credit offers generated by AI assistants for instance, while avoiding inappropriate recommendations from old data.

Data Quality Management

Equally important is managing data quality - the discipline of ensuring datasets meet fitness needs of target users and systems. For generative models, specific quality KPIs are defined covering dimensions like accuracy, consistency, completeness relative to model requirements. Data governance introduces control points in pipelines to measure against thresholds and trigger alerts for remediation. This expands from passive reporting to active data issue resolution closing the loop.

Metadata Governance

Like highways enable transportation, metadata provides the data mapping and meaning that allows discovery and understanding across distributed environments. Metadata governance maintains this navigational layer connecting data producers, stewards, engineers and scientists to speed accessibility while preventing fragmentation. Tagging schemas developed for generative model inputs get propagated across tooling to accelerate data preparation. Glossaries clarify terminology reducing confusion for end users while enabling metadata-driven automation in MLOps pipelines.

Data Ethics and Regulations

An often overlooked aspect of data governance for generative AI is managing ethics risk and external regulations. Generative models can perpetrate harm through biases, inappropriate content and more. Data governance helps embed ethics into model data supply chains through privacy preserving approaches like tokenization, noise infusion and bias detection. Compliance to growing regulations like GDPR is operationalized through governance protocols on securing data , managing subject consent and enabling the right to erasure. This prevents severe regulatory, reputation and reliability risks that outweigh efficiency gains.

In summary, Generative AI needs flexible yet governed data. Data governance provides the foundation and guardrails that enable business scale by accelerating data ready for models while ensuring its suitability and integrity for positive impact.

Building Relevant Data Capabilities

Realizing the full potential of generative AI, powered by strategic frameworks, robust architectures, and comprehensive governance, hinges on the cultivation of specialized data capabilities. Leading organizations are channeling their efforts into enhancing three critical areas to build these capabilities: nurturing advanced analytics talent, deploying self-service data access platforms, and adopting agile team constructs.

Advanced Analytics Talent

The advent of generative algorithms has escalated the demands for data across volume, variety, and velocity. Addressing these demands necessitates a pool of advanced analytics talent, including:

- **Data Engineers** tasked with consolidating, transforming, and pipelining vast, streaming datasets into dependable inputs for models, often wielding expertise in distributed data processing frameworks such as Apache Spark.

- **Data Architects** who craft the enterprise's data blueprints, capitalizing on cloud-native storage and analytics services, and setting standards for data interchange formats and governance protocols.

- **Data Scientists** engaged in preparing multi-structured data, including sensor streams, images, and text, through techniques like statistical imputation and feature extraction, while providing insights on model feedback mechanisms.

- **MLOps Engineers** focused on operationalizing the model lifecycle, monitoring data collection, model building, evaluation, and tuning, and maintaining oversight of data and model metrics.

- **Prompt Engineers**, specialists in designing and optimizing prompts for language models, crucial for tailoring generative AI outputs to specific tasks and ensuring the relevance and quality of generated content.

Self-Service Data Access Platforms

These platforms empower a broader user base with self-service access to prepared datasets, eliminating bottlenecks in model development. They enhance visibility into model and data metrics, fostering continuous improvement through features like:

- **Data Discovery**, which allows users to navigate curated datasets via metadata catalogues and engage in interactive querying without deep knowledge of the underlying infrastructure.

- **Model Development Environments**, equipped with popular open-source libraries such as TensorFlow and PyTorch, support rapid experimentation by citizen data scientists, paving the way for IT deployment.

- **MLOps Platforms**, offering tools for model and metric monitoring, automated CI/CD pipeline executions, and alerts for model retraining prompted by data drift.

Agile Team Constructs

Inspired by successful software development practices, agile team models like pods foster close collaboration among data engineers, scientists, and ML engineers—essential for the success of generative AI initiatives. These small, integrated teams can swiftly iterate on the data needs of targeted model applications, leveraging platforms and modular architecture to expedite experiments and the realization of value.

In essence, building relevant data capabilities forms the essential infrastructure around strategies, architectures, and governance frameworks, propelling the adoption of generative AI and enhancing its impact on business.

AI-Native Thinking in Data Strategy and Data Management

In the era of rapid technological advancement, AI-Native thinking has emerged as a cornerstone for organizations aspiring to leverage artificial intelligence to its fullest potential. This approach fundamentally reshapes traditional concepts of data strategy and data management, placing AI at the core of business operations and decision-making processes. Here's how AI-Native thinking redefines these areas:

- **Data as a Core Business Driver**: For AI-Native organizations, data transcends its conventional role as a support element and becomes a central business driver. This paradigm shift requires a data strategy that is intricately aligned with the company's core business objectives, recognizing data as a crucial asset that drives value through enhanced efficiencies, customer experiences, and new revenue models.

- **Ethical Data Monetization in the AI Era**: In AI-Native entities, data monetization is approached not just as a revenue channel but as a strategic and ethical practice. These organizations leverage their data to enhance products or services, create new data-driven offerings, or responsibly sell insights, always prioritizing customer privacy and adhering to ethical standards.

- **Data Governance Tailored for AI**: AI-Native thinking demands comprehensive data governance that transcends traditional boundaries. This includes rigorous policies and practices for data quality, security, and privacy, but also extends to ensuring the ethical use of AI and compliance with evolving regulations like GDPR and CCPA. In this framework, governance is a dynamic, evolving process that adapts to the changing landscapes of AI and data regulation.

- **Seamless AI and Analytics Integration**: In AI-Native strategies, the integration of AI with data management systems is not an afterthought but a foundational aspect. This integration supports advanced and predictive analytics, enabling organizations to derive profound insights and inform strategic decisions. The emphasis is on high-quality, well-curated data that fuels effective AI algorithms.

- **Data Architectures Built for AI Scalability**: Recognizing the dynamic nature of AI and its data needs, AI-Native organizations adopt data architectures that are inherently scalable and flexible. These architectures, whether cloud-based, data lakes, or hybrid systems, are designed to handle large and evolving datasets, ensuring seamless integration with emerging AI technologies.

- **Fostering an AI-First Culture**: AI-Native thinking involves cultivating a culture where data and AI are at the heart of every decision and operation. This requires a commitment to enhancing data literacy across the organization, empowering employees to leverage data and AI insights in their roles. It's about building an environment where data-driven decision-making is the norm.

- **Emphasizing Real-Time Data and Responsiveness**: In an AI-Native framework, the capacity for real-time data processing and analytics is vital. This capability allows organizations to respond swiftly to market changes and customer needs, providing a competitive edge in today's fast-paced business environment.

- **Upholding Data Privacy and Ethical AI**: AI-Native organizations place a high premium on data privacy and the ethical implications of AI. This involves ensuring that AI algorithms are transparent, fair, and unbiased, and that data practices respect user privacy. Ethical AI is seen not just as a compliance requirement but as a critical aspect of building and maintaining trust.

- **Robust Data Security as a Foundation**: With the increasing reliance on data, AI-Native entities prioritize advanced data security measures. This includes not only protecting against breaches and unauthorized access but also regularly auditing and updating security practices to stay ahead of emerging threats.

- **Commitment to Continuous AI Evolution**: Lastly, an AI-Native approach to data strategy and management is characterized by a continuous improvement ethos. Regular reviews and updates of data strategies and technologies are essential to keep pace with the rapidly evolving AI and data landscapes.

In summary, AI-Native thinking in data strategy and data management represents a transformative approach, where AI is not just an added tool but a fundamental aspect of how data is viewed, managed, and utilized. It demands a holistic reconsideration of data's role in the organization, emphasizing ethical practices, real-time responsiveness, continuous evolution, and a deep integration of AI into every facet of the business.

Conclusion

Generative AI ushers in a new era of possibilities for enterprises to transform products, engage customers and unlock growth opportunities in imaginative new ways. Behind the incredible potential of writing algorithms and content creating bots lies data - their lifeblood and limiting factor.

Without careful planning and sustained investment in data readiness, generative AI initiatives risk stalled progress, with scattered pilots failing to catalyze enterprise value at scale. Success lies in recognizing data readiness as a strategic capability vital to fueling generative AI and calls for executive leadership commitment.

This starts with developing an AI data strategy that audits current assets and blueprints a roadmap prioritizing high-impact enrichments. Modern, cloud-accelerated data architecture provides the secure, cost-efficient infrastructure to ingest, refine and serve torrents of multi-structured data. Data governance introduces the guardrails for speed, scale and ethics adherence.

And finally, talent and team models transform blueprints to business impact. Advanced data engineers, scientists and ML practitioners closely partner in agile teams to continuously channel governed data into optimized model pipelines - powering the flywheel of improvement through validated synthetic content, products and experiences.

With careful orchestration of people, process and technology, companies can effectively harness the exponential power of generative AI to shape futures previously unimagined. The data journey demands vigilance and vision, but unlocks immense opportunity for pioneers able to purposefully prepare, architect and govern data as a strategic capability lifting all boats in a rising AI tide.

Part Four: Embarking on Your AI Transformation Journey

ArgoLong Publishing

Chapter 15

Jump-Start Your AI Initiative

"A journey of a thousand miles begins with a single step. " — *Lao Tzu*

I n the swiftly evolving landscape of technology, Artificial Intelligence (AI) emerges as a transformative force, redefining the competitive dynamics across industries. For contemporary business leaders, the pertinent question has shifted from whether to integrate AI into their operations to when and how to do so with maximum efficacy. The inception of an AI strategy brims with complexities and challenges. This chapter is designed to shepherd leaders through the initial phase of their AI journey, providing actionable insights to surmount prevalent obstacles and secure a triumphant commencement.

The promise of AI to boost operational efficiency, drive innovation, and enhance decision-making capabilities is immense. Yet, embarking on the AI path often presents a daunting challenge for many organizations: determining the starting point. Leaders frequently ponder, **"How do I initiate an AI strategy effectively?"** Tackling this question necessitates a nuanced understanding of the challenges associated with launching an AI strategy and devising strategic measures to navigate them successfully.

Among the primary challenges encountered are:

- **The Chicken or Egg Dilemma**: A common conundrum for leaders is deciding whether to initially invest in AI technologies or to formulate compelling business cases. The task of developing influential business cases is challenging without firsthand experience with AI technologies.

- **Approach to Adoption**: Leaders are faced with the choice of adopting a top-down approach, where directives and vision stem from the top echelons of leadership, a bottom-up approach that leverages grassroots innovation, or a blend of both. This decision is crucial to ensure the seamless integration of AI within their organizations.

- **Budget Considerations**: Contrary to the widespread notion, initiating AI projects does not necessarily require a significant financial outlay from the outset. An adaptive, iterative approach permits a more measured and budget-friendly entry into the realm of AI.

This chapter aims to clarify these initial challenges, guiding leaders through the essential steps to commence their AI adoption journey. By understanding the intricacies of initiating an AI initiative, leaders can establish a solid foundation for a journey that not only yields substantial returns but also provides strategic advantages in the long term.

The Chicken or Egg Problem: First-hand Experience vs. Building Business Cases

In the formative phases of adopting artificial intelligence (AI), business leaders often grapple with a pivotal decision: whether to embark on AI projects to garner first-hand experience or to meticulously craft business cases as a precursor to investing in AI. This conundrum, colloquially known as the "Chicken or Egg" dilemma, stands as a formidable challenge in laying the groundwork for AI strategies.

The Case for First-Hand Experience

- **Understanding AI Capabilities and Limitations**: Direct involvement with AI projects offers leaders a real-world understanding of what AI can and cannot do. This hands-on experience is invaluable in forming realistic expectations and strategies.

- **Innovation and Discovery**: Engaging with AI technologies can lead to unexpected discoveries and innovative applications specific to the organization's needs, which might not be evident when only building theoretical business cases.

- **Building Internal Advocacy**: First-hand success stories, even from small-scale projects, can build enthusiasm and support for AI within the organization, creating a stronger foundation for larger initiatives.

The Case for Building Business Cases

- **Strategic Alignment**: Comprehensive business cases guarantee that AI initiatives are congruent with the organization's overarching strategic objectives, mitigating the risk of misallocation of resources.

- **Risk Assessment**: The process of crafting business cases entails a rigorous analysis of potential risks and benefits, facilitating informed decision-making.

- **Resource Allocation**: A well-articulated business case can provide the necessary justification for the allocation of resources, ensuring that AI projects receive adequate funding and support.

Strategies for Overcoming the Dilemma

- **Integrating Both Approaches**

- : A judicious approach may entail initiating small-scale AI endeavors to acquire insights and experience, which in turn, can enrich more elaborate business cases for more extensive initiatives.

- **Embracing Iterative Progression**: An iterative strategy, where AI projects commence on a modest scale and expand based on success and accrued knowledge, can amalgamate the practical benefits of firsthand experience with the prudence of strategic planning.

- **Inclusive Stakeholder Engagement**: Soliciting input from a broad spectrum of stakeholders during both exploratory efforts and the development of business cases ensures a diversity of perspectives and consolidates support.

The decision between diving into AI and building robust business cases first is not a straightforward one and varies based on the organization's context, culture, and resources. A pragmatic approach might involve a combination of both – starting with exploratory AI projects to build understanding and advocacy, and then developing detailed business cases as these projects evolve and require more significant investment. This approach ensures a balance between strategic alignment and the innovation that comes from first-hand experience with AI.

AI Adoption Strategies: Top-Down, Bottom-Up, and Hybrid

The pathway an organization chooses to integrate artificial intelligence (AI) plays a pivotal role in the success of its AI endeavors. Three primary strategies are at the forefront: top-down, bottom-up, and a hybrid model, each with its distinct advantages and obstacles. The decision on which path to follow largely hinges on the organization's culture, structural dynamics, and overarching strategic goals.

Strategy	Characteristics	Advantages	Challenges	Ideal Use Case
Top-Down Approach	Driven by organization's leadership, setting strategic AI objectives at the executive level.	Ensures alignment with organizational goals, clear direction, and resource allocation.	May overlook grassroots innovation and insights from day-to-day operations.	Organizations where strategic alignment and control are paramount, especially larger or regulated industries.
Bottom-Up Approach	Initiatives start at the grassroots level, driven by employees or departments identifying AI opportunities.	Encourages innovation, hands-on experimentation, and tailored AI solutions.	Risk of strategic misalignment and scaling difficulties.	Organizations with a strong culture of innovation and empowerment, effective in dynamic environments.
Hybrid Approach	Combines top-down and bottom-up approaches with strategic direction from leadership and innovation from all levels.	Balances strategic alignment with innovation, encouraging organization-wide engagement.	Requires effective communication, collaboration, and alignment mechanisms.	Most organizations seeking balance between strategy and innovation, where leadership is open to ideas from all levels.

Table 15-1: The Comparison of Three AI Adoption Strategies

Top-Down Approach: Strategic Direction from the Helm

- **Characteristics**: This approach is driven by the organization's leadership. It involves setting strategic AI objectives at the executive level and cascading them down through the organization.

- **Advantages**: Ensures alignment with broader organizational goals and provides a clear direction for AI initiatives. It also helps in securing necessary resources and fostering organization-wide commitment to AI projects.

- **Challenges**: May lack grassroots innovation and may not fully leverage the insights and knowledge of employees who are closer to the day-to-day operations.

- **When to Use**: This approach is ideal for organizations where strategic alignment and centralized control are paramount. It's best suited for larger organizations or those in highly regulated industries where a unified AI vision and compliance are critical.

Bottom-Up Approach: Empowering Grassroots Innovation

- **Characteristics**: In this approach, AI initiatives often start at the grassroots level, driven by employees or specific departments identifying opportunities for AI applications.

- **Advantages**: Encourages innovation and allows for practical, hands-on experimentation with AI. It can also lead to more tailored AI solutions that directly address specific operational challenges.

- **Challenges**: Risks include a lack of strategic alignment and potential difficulties in scaling small, disparate projects into organization-wide solutions.

- **When to Use**: This approach works well in organizations with a strong culture of innovation and empowerment at the employee level. It is particularly effective in dynamic environments where operational teams may have a clearer understanding of where AI can be applied to solve immediate challenges.

Hybrid Approach: Merging the Best of Both Worlds

- **Characteristics**: This strategy weaves together the top-down and bottom-up approaches, with leadership defining strategic AI goals and allocating resources, while also encouraging widespread innovation and experimentation among all employees.

- **Advantages**: Balances strategic alignment with grassroots innovation. It encourages organization-wide engagement in AI initiatives, potentially leading to more effective and sustainable AI solutions.

- **Challenges**: Requires effective communication and collaboration across different levels of the organization, and may need mechanisms to ensure that innovation is aligned with strategic goals.

- **When to Use**: The hybrid approach is suitable for most organizations seeking a balance between strategic alignment and innovation. It works well in environments where leadership is open to ideas from all levels of the organization and where there is a need to ensure that AI initiatives are both strategically relevant and practically applicable.

Selecting the appropriate adoption strategy requires a thoughtful analysis of the organization's culture, structure, strategic aspirations, and the unique demands of its AI initiatives. A deep understanding of these elements will enable leaders to choose the most suitable approach for embarking on their AI adoption journey.

Starting with a Limited Budget: The Agile Way

The Agile methodology, characterized by its iterative and incremental approach, is particularly well-suited for managing AI projects, especially when budget constraints are a factor. This approach allows organizations to start small, learn quickly, and adapt as they go, reducing the need for substantial upfront investments.

Key Elements of the Agile Approach in AI

- **Small-Scale Initiatives**: Begin with small-scale AI projects that require minimal investment. These can be pilot programs or proof-of-concept initiatives that focus on addressing specific, manageable problems.

- **Iterative Development**: AI projects should be developed in short, iterative cycles. This allows for continuous feedback and adjustments, reducing the risk of large-scale failures and enabling more efficient allocation of resources.

- **Flexibility and Adaptability**: Agile methodologies emphasize flexibility and adaptability. As AI projects evolve, the ability to pivot or change direction based on learnings and results is crucial, especially when working with limited resources.

- **Stakeholder Engagement**: Involving various stakeholders, including end-users and team members, in the development process ensures that AI solutions are aligned with user needs and organizational goals.

- **Measuring Progress and Value**: Regularly measure the progress and value of AI initiatives against set goals and objectives. This helps in justifying continued investment in the project and in making informed decisions about scaling or altering the project scope.

When to Use the Agile Approach

- **Complex, Uncertain Projects**: AI projects often involve a high degree of complexity and uncertainty. The Agile approach is well-suited for such environments where requirements and solutions evolve through the collaborative effort of cross-functional teams.

- **Limited Budget Scenarios**: When funding is a constraint, the Agile methodology allows organizations to make smaller, incremental investments, assess their impact, and then decide on further investments based on tangible results.

- **Innovation-Focused Environments**: For organizations looking to foster innovation and quick adaptation to changing market needs, Agile offers a framework that supports rapid experimentation and learning.

Starting AI initiatives with a limited budget is not only feasible but can also be strategically advantageous using the Agile approach. By starting small, focusing on specific problems, and adopting an iterative development cycle, organizations can effectively manage their resources while still exploring the benefits of AI. This approach not only mitigates financial risks but also encourages a culture of innovation and continuous improvement, essential for long-term success in AI endeavors.

Identifying and Empowering AI Enthusiasts

When initiating an AI project, it's crucial to recognize that individuals within an organization often fall into one of three categories regarding AI: passionate, opposed, and neutral observers.

- **Passionate Advocates**: These individuals are enthusiastic about AI and understand its potential. They are often early adopters of new technologies and are keen to explore AI's possibilities.

- **Opponents**: There are also those who are skeptical or resistant to AI, often due to concerns about job displacement, complexity, or a lack of understanding of AI's benefits.

- **Neutral Observers**: A large group typically watches from the sidelines. They may be open to AI but haven't formed a strong opinion either for or against it.

Leveraging Passionate Advocates as Change Agents

In the realm of AI initiatives, passionate advocates play a pivotal role. These individuals are not just supporters of AI; they often embody a forward-thinking mindset and a deep belief in the transformative power of AI. Their passion for AI makes them ideal candidates to act as change agents within an organization.

The Role of Passionate Advocates

Passionate advocates bring more than just enthusiasm to the table. They typically possess or are keen to develop a deeper understanding of AI technologies and applications. This knowledge, combined with their zeal, positions them uniquely to lead and inspire others. They can effectively communicate the potential of AI, address misconceptions, and showcase tangible benefits, thereby demystifying AI for their colleagues.

Turning Enthusiasm into Leadership

These advocates can be nurtured into leadership roles within AI projects. Their first-hand experience with AI tools and technologies, coupled with an innate curiosity, allows them to explore innovative applications of AI within the organization. They can spearhead pilot projects, lead exploration teams, and be the driving force behind AI adoption.

The Impact of Change Agents

When passionate advocates are positioned as change agents, they can significantly impact the organization's AI journey. They act as catalysts, accelerating AI adoption, and creating a ripple effect throughout the organization. Their enthusiasm can be infectious, encouraging others to explore and embrace AI technologies.

Overcoming Resistance with Advocacy

One of the critical roles of these change agents is to bridge the gap between AI supporters and skeptics within the organization. Through their advocacy, they can alleviate fears and concerns about AI, helping to transform resistance into acceptance and even enthusiasm. By sharing their experiences and success stories, they can illustrate how AI can be a tool for enhancement rather than a threat, thereby changing perceptions across the organization.

Supporting Change Agents

For organizations to fully leverage these passionate advocates, it's essential to support them adequately. This support can come in the form of providing access to resources, training, and platforms where they can share their knowledge and insights. Recognizing and valuing their contributions is also crucial in maintaining their enthusiasm and commitment.

In summary, passionate advocates can be the linchpins in an organization's AI strategy. By harnessing their enthusiasm and knowledge, and elevating them to change agent roles, organizations can effectively drive their AI initiatives forward, fostering a culture that embraces innovation and technological advancement.

Volunteer-Based Approach for AI Team Assembly

A volunteer-based approach for team assembly in AI initiatives is a strategy that taps into the intrinsic motivation and genuine interest of employees. Rather than assigning team members in a top-down manner, this approach invites volunteers who are enthusiastic about participating in AI projects. This methodology can be particularly effective for jump-starting AI initiatives for several reasons.

Firstly, a volunteer-based team is often more engaged and motivated. The individuals who choose to be part of the AI initiative are likely to be those who have a natural interest in the technology or see its potential to make a positive impact in their work. Their self-motivation can drive the project forward with energy and commitment that might not be as strong in a team formed solely through assignments.

Secondly, this approach can tap into a diverse range of talents and perspectives. Volunteers may come from various departments and levels within the organization, bringing a rich mix of insights, experiences, and skills. This diversity can be particularly beneficial in AI projects, which often benefit from multidisciplinary approaches and creative problem-solving.

Moreover, forming a team based on voluntary participation also fosters a sense of ownership and empowerment among the team members. When individuals choose to be part of a project, they are more likely to feel invested in its success and contribute proactively. This sense of ownership can enhance the quality of work and encourage team members to go above and beyond in their roles.

However, it's important to manage such teams effectively to ensure that the enthusiasm translates into productive outcomes. Clear goals and expectations should be set, and adequate support and resources should be provided. Leadership should also ensure that the volunteer-based team has access to necessary training and mentoring, especially if the volunteers are new to AI.

Additionally, recognizing and celebrating the contributions of these volunteer teams is crucial. Acknowledgment of their efforts not only boosts morale but also signals to the broader organization the value and impact of engaging with AI initiatives.

In summary, a volunteer-based approach for assembling an AI team can catalyze the initiative with genuine enthusiasm and diverse insights. It leverages the inherent interests and talents within the organization, fostering an environment of innovation and collaboration that is well-suited for the exploratory and dynamic nature of AI projects.

Training and Supporting the Team

Training and supporting the team is a critical component in ensuring the success of AI initiatives. Proper training equips team members with the necessary skills and knowledge, while support ensures they have the resources and environment conducive to innovation and effective problem-solving.

Training Aspects

- **Skills Development**: The first step is to identify the specific skills required for your AI projects. This could range from prompt engineering, data science and machine learning to AI ethics and project management. Training programs should be tailored to fill these skill gaps.

- **Continuous Learning**: AI is a rapidly evolving field. Encourage a culture of continuous learning where team members are motivated to stay updated with the latest developments and technologies in AI. This can be facilitated through workshops, webinars, and online courses.

- **Customized Training Programs**: Consider the varying skill levels and learning styles within your team. Some members might benefit from in-depth technical courses, while others might need more foundational training. Customized training programs can cater to these diverse needs.

- **Real-World Applications**: Incorporate practical, hands-on training sessions where team members can apply their learning to real-world scenarios. This approach helps in cementing theoretical knowledge through practical experience.

Supportive Environment

- **Collaborative Workspace**: Foster a collaborative environment that encourages sharing of ideas and knowledge. This could be through regular team meetings, brainstorming sessions, or collaborative platforms for knowledge exchange.

- **Mentorship and Guidance**: Pairing team members with mentors who have experience in AI can be extremely beneficial. Mentors can provide guidance, share insights, and help navigate complex challenges.

- **Resource Availability**: Ensure that the team has access to necessary resources, including software tools, datasets, and computing power. Having the right tools at their disposal is crucial for effective AI development.

- **Psychological Safety**: Create an atmosphere where team members feel safe to experiment, take risks, and even fail. This kind of psychological safety is key to fostering innovation and creativity.

- **Recognition and Reward**: Recognize and reward the efforts and achievements of your AI team. This not only boosts morale but also reinforces the value of their work to the organization.

Training and supporting an AI team is not just about technical skill development; it's also about creating an environment that nurtures continuous learning, collaboration, innovation, and resilience. By investing in comprehensive training and supportive measures, organizations can significantly enhance the effectiveness and satisfaction of their AI teams, ultimately contributing to the success of their AI initiatives.

Addressing the Other Camps

Addressing the other camps – the opponents and the neutral observers – is a crucial aspect of successfully implementing an AI initiative. While the passionate advocates are important for driving the initiative, engaging with those who are skeptical or indifferent towards AI is essential for holistic adoption and success.

Engaging Opponents

Opponents of AI, often skeptical due to concerns about job displacement, ethical implications, or a lack of understanding of AI's benefits, require a tailored approach:

- **Education and Communication**: One of the most effective ways to address concerns is through education. Providing clear, jargon-free information about what AI is and what it isn't can demystify the technology and alleviate fears.

- **Highlighting Benefits**: Showcase how AI can augment human capabilities rather than replace them. Demonstrating AI's potential to take over mundane tasks, thereby freeing employees for more creative and meaningful work, can be a persuasive argument.

- **Inclusive Decision-Making**: Involving skeptics in the decision-making process can give them a sense of control and ownership over the technology. Their insights can also be valuable in identifying potential pitfalls and ethical considerations.

- **Success Stories**: Sharing success stories, especially from within the organization or similar industries, can help in changing perceptions. Real-world examples where AI has led to job creation, improved efficiency, or other positive outcomes can be influential.

Engaging Neutral Observers

Neutral observers, while not overtly resistant to AI, may lack sufficient information or interest to form an opinion. Engaging this group is also important:

- **Encouraging Curiosity**: Sparking curiosity can be done through workshops, seminars, or informal discussions. The goal is to turn their neutrality into interest or even enthusiasm.

- **Peer Influence**: Leveraging the influence of passionate advocates can be effective. Observers are more likely to be swayed by the positive experiences and testimonials of their colleagues.

- **Opportunities for Involvement**: Offering opportunities for observers to get involved in AI projects at a comfortable level can help convert their passive stance into active participation.

- **Regular Updates**: Keeping everyone in the loop with regular updates about the progress and achievements of AI initiatives can help in maintaining an interest and gradually building support.

Successfully navigating the landscape of varying attitudes towards AI in an organization requires a multifaceted approach. While leveraging the energy of the passionate advocates is crucial, equally important is the need to educate, engage, and involve the skeptics and the observers. Through a combination of communication, education, inclusion, and showcasing of benefits, organizations can foster a more widespread acceptance and enthusiasm for AI, paving the way for a successful and comprehensive AI adoption.

Setting Sail on Your AI Journey

Embarking on an AI initiative is a journey that requires careful consideration, strategic planning, and an inclusive approach. This chapter has navigated through the essential steps and considerations for jump-starting an AI initiative, highlighting the importance of addressing the initial challenges and leveraging the right strategies and people.

The journey begins with resolving the 'Chicken or Egg' dilemma, where organizations must balance the need for hands-on experience with the development of strong business cases. Adopting a top-down, bottom-up, or hybrid approach is crucial in aligning AI initiatives with organizational goals while fostering innovation. Starting with a limited budget, the Agile methodology proves effective, allowing for incremental growth and learning.

Central to the success of any AI initiative is the people who drive it. Leveraging passionate advocates as change agents, adopting a volunteer-based approach for team assembly, and effectively addressing the concerns of skeptics and neutral observers are key steps in building a robust AI team. This team, supported by the right training and resources, becomes the cornerstone of AI adoption and innovation within the organization.

In conclusion, while the path to initiating and successfully implementing AI can be complex, the benefits far outweigh the risks. By starting small, embracing a culture of learning and adaptation, and engaging the right people, organizations can harness the transformative power of AI. This journey is not just about technology; it's about cultivating a mindset and environment that embraces change and innovation. As we stand at the cusp of a new era of technological advancement, organizations have the opportunity to lead the charge, exploring uncharted territories of AI, and reaping its vast rewards. The key is to take that first step with confidence and clarity, ensuring a future where AI is an integral and value-adding part of the business landscape.

Chapter 16

First Thing First: Mastering the Prompt Engineering Skill

"First things first; take care of what can be done now before worrying too long over what might never be." — Robert Jordan

E mbarking on the journey into the realm of Generative AI, organizations are often brimming with enthusiasm and visions of transformative potential. They foresee leveraging AI to revolutionize their operations, spur innovation, and carve out competitive edges in rapidly evolving markets. Yet, as they stand on the precipice of this technological leap, a critical challenge emerges—one that is as foundational as it is frequently underestimated: **How to write effective prompts?**

This challenge, seemingly simple at first glance, quickly unravels into a complex obstacle that can significantly impact the trajectory of AI initiatives. The crux of the matter lies not in the lack of ideas or the absence of technological infrastructure but in the art and science of communicating those ideas to the AI. The most common stumbling block organizations encounter is crafting prompts that accurately convey their needs and intentions to these advanced systems. Effective prompt writing is not just about what is being asked but how it's being asked, ensuring that the AI can grasp the context, nuances, and specific goals of the request.

The importance of this challenge cannot be overstated. A well-crafted prompt can unlock the full capabilities of Generative AI, leading to outputs that are not only relevant and precise but also innovative and insightful. Conversely, a poorly constructed prompt can result in outputs that miss the mark, wasting valuable time and resources, and potentially leading to frustration and disillusionment with AI technologies.

This chapter delves into the essence of this initial hurdle, exploring the intricacies of prompt engineering. We will uncover the full spectrum of Generative AI outputs, illustrating the vast potential that lies within effective communication with

AI systems. From understanding the power of well-constructed prompts in AI interactions to mastering the skill of prompt engineering, this chapter provides a comprehensive guide for organizations and individuals alike.

Through the lens of prompt engineering, we will explore whether this skill is a niche profession or an essential competency for the future. With practical insights and case studies, this chapter aims to equip readers with the knowledge and tools to master the art of prompt engineering, transforming the common challenge of writing effective prompts into a stepping stone for success in the AI-driven world.

Unveiling the Full Spectrum of Generative AI Outputs

As we navigate through the transformative era of artificial intelligence, the prowess of generative AI becomes increasingly evident, reshaping our approach to creativity, problem-solving, and decision-making across a myriad of domains. This evolution underscores the essence of generative AI: its power lies inherently in the diversity and sophistication of its outputs. These outputs not only span across textual and visual content but also extend into code generation, design conceptualization, data analytics, strategy development, and beyond, illustrating the expansive capabilities of generative AI systems. Understanding and leveraging these outputs is pivotal for maximizing the potential of AI in both creative and technical fields.

The Foundation of Creativity: Textual and Visual Outputs

Generative AI's ability to produce rich, diverse textual content such as articles, reports, essays, and creative writing marks the beginning of its transformative impact. This versatility extends to the visual domain, where AI systems, through techniques like Generative Adversarial Networks (GANs), generate original artworks, photorealistic images, and design mockups. These capabilities provide an endless wellspring of creativity and innovation for writers, marketers, artists, and designers, enabling the creation of content that was once beyond the reach of traditional methods.

The Harmony of Sound: Audio and Music Composition

The generative power of AI transcends visual creativity to encompass the auditory domain, offering unique compositions, sound effects, and voiceovers. This aspect of AI is revolutionizing the music industry, film production, and multimedia content creation, allowing for the customization of audio content to fit specific narratives or emotional tones.

Bringing Ideas to Life: Video, Animation, and 3D Modeling

Generative AI's foray into video generation, animation, and 3D modeling is crafting new dimensions for storytelling and environmental design. From short clips to detailed virtual environments, AI facilitates the visualization of complex ideas and stories, offering tools that allow creators to transcend traditional constraints.

The New Frontier: Interactive and Dynamic Content

Beyond static creations, generative AI excels in producing interactive and dynamic content, adapting in real-time to user interactions. This capability is pivotal in creating personalized experiences in education, gaming, and digital platforms, highlighting the AI's role in crafting more engaging and effective user experiences.

Innovating Problem-Solving: Code and Test Case Generation

Generative AI's impact extends into the realm of software development, where it accelerates the creation of code and test cases, enhancing productivity and software robustness. This application underscores the AI's role in streamlining complex workflows and improving the quality of technological products.

Envisioning the Future: Idea and Design Conceptualization

In design and ideation, generative AI acts as a catalyst for innovation, providing a breadth of ideas and visualizations that fuel creativity and significantly reduce the time from conception to prototype. This capacity for rapid ideation and conceptualization is revolutionizing product design, architecture, and creative industries.

Deciphering Complexity: Data Analytics and Insights

Generative AI's ability to analyze vast datasets and generate predictive models offers unprecedented insights across finance, healthcare, and marketing. This analytical power enables more informed decision-making, driven by AI's capacity to uncover patterns and forecast trends beyond human capability.

Charting Paths Forward: Strategy and Method Development

In strategic planning and methodological development, generative AI proposes optimized processes and strategies, demonstrating its utility in business optimization, scientific research, and beyond. This strategic application showcases AI's role in enhancing efficiency and effectiveness across various sectors.

In summary, the core strength of generative AI lies in the breadth and depth of its outputs, a testament to its potential to revolutionize industries, enhance creativity, and streamline processes. By harnessing the full spectrum of capabilities offered by generative AI, individuals and organizations unlock new avenues for innovation, optimization, and creation. As we look forward to the future, the ongoing advancement of generative AI promises to expand the horizons of what is possible, heralding a new era of technological empowerment and creative exploration.

The Secret Sauce to Mastery: Effective Prompts in AI Interactions

In the vast and evolving landscape of artificial intelligence, the ability to craft and refine prompts emerges as the linchpin for successful human-AI communication. This chapter explores the intricate balance between art and science required to create effective prompts, a skill indispensable for unleashing the full capabilities of AI technologies, including OpenAI ChatGPT, Google Bard, and Anthropic Claude. Crafting precise and impactful prompts is likened to discovering a secret sauce—a special ingredient that elevates the quality of AI-generated outputs from ordinary to exceptional.

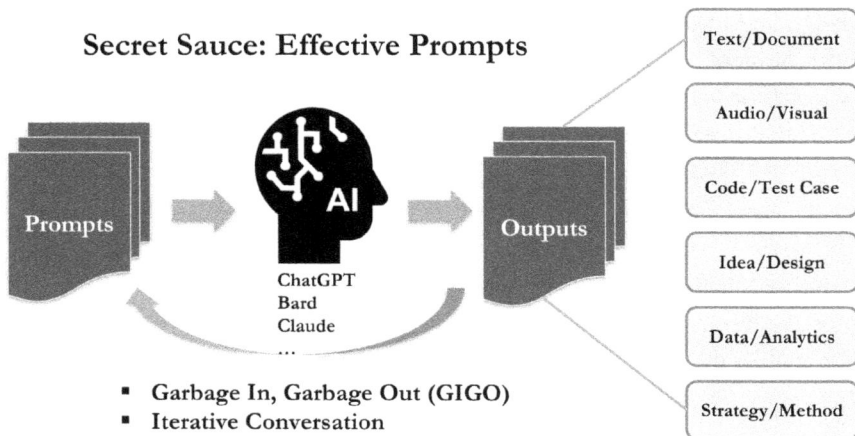

Figure 16-1: Effective Prompts to Generate a Wide Variety of Outputs

The Principle of "Garbage In, Garbage Out" (GIGO)

At the core of computing and AI is the principle of "Garbage In, Garbage Out" (GIGO), which emphasizes the critical importance of quality input. In prompt engineering, this principle manifests in the careful construction of prompts that are clear, concise, and customized to the AI's specific strengths and weaknesses. The quality of input directly influences the AI's ability to engage in iterative conversations, tackle complex problems, and generate innovative solutions, marking the first step toward effective AI utilization.

The Diverse Landscape of Prompts

The realm of effective prompts extends beyond mere text-based interactions, encompassing a variety of formats such as audio/visual cues, code/test cases, creative ideas/design concepts, data/analytics inquiries, and strategic/methodological explanations. This diversity highlights the versatility and critical nature of prompt engineering, establishing it as a fundamental skill across different fields and mediums, essential for engaging with AI in sophisticated and productive manners.

The Dynamics of Iterative Conversation

The concept of iterative conversation underlines the dynamic essence of human-AI interaction, where effective prompts catalyze a reciprocal exchange. This process allows each AI response to refine the user's understanding and subsequent inquiries, leading to more precise and meaningful AI outputs. It signifies a departure from static commands to an ongoing dialogue, empowering users to delve into complex ideas, resolve uncertainties, and achieve more comprehensive and innovative outcomes.

Navigating the Nuanced Terrain of Human and AI Interactions

Human-AI Communication ≠ Human-Human Communication

Aspect	Human-Human	Human-AI
Nature & Fluidity	Organic, emotional, fluid	Programmatic, pattern-based, less fluid
Context & Interaction	Implicit understanding, bidirectional	Explicit context needed, unidirectional
Input & Assumptions	Vague, inferred based on shared knowledge	Precise, requires background/context in prompts
Memory & Error Correction	Selective memory, clarifications	No genuine memory, re-prompting for error correction
Feedback Mechanisms	Verbal/non-verbal cues, questions	Refining prompts based on AI responses

Table 16-1: The Comparison of Human-AI Communication and Human-Human Communication

While human-human and human-AI communications both aim for clarity and understanding, their paths diverge in execution and essence, yet they share some common ground:

- **Clarity and Conciseness**: Both communication forms prize clarity and brevity—whether it's articulating thoughts between people or instructing an AI.

- **Feedback for Growth**: Feedback acts as a nurturing tool in human relationships and a fine-tuning mechanism in AI interactions, essential for progression and enhancement.

- **Context is King**: In both scenarios, context enriches communication, setting the scene for deeper understanding and more relevant interactions.

However, distinct differences underscore the unique challenges of prompt engineering:

- **Emotional Intelligence**: Crucial for human interaction and nuanced understanding, this element is nascent or entirely absent in AI communications.

- **The Flow of Conversation**: Human conversations naturally ebb and flow, reflecting the complexity of human thought, whereas AI interactions tend to follow a more structured, direct path.

- **Adaptability**: Humans instinctively adjust to shifts in dialogue, akin to chameleons changing color. AI, however, operates within the confines of its training, exhibiting a limited capacity for deviation from learned responses—a characteristic limitation of current AI technologies.

In bridging these divergences, the art of crafting effective prompts becomes not just a technical skill but a nuanced craft, essential for anyone looking to navigate the future landscape of AI with confidence and creativity.

Prompt Engineering: Unlocking AI's Potential

Definition: Prompt Engineering is the art and science of crafting, evaluating, and refining prompts that direct AI models to generate desired outputs. It combines deep technical understanding of models' inner workings and natural language processing with linguistic expertise and creativity, aiming to optimize human-AI interactions across various applications.

Prompt engineering emerges as a vital discipline aimed at mastering the art and science of crafting effective prompts to elicit desired outcomes from AI systems. This field transcends mere question-asking; it embodies the strategic formulation of queries that enable AI systems to grasp and respond with precision. As artificial intelligence weaves itself more intricately into the fabric of our everyday lives, the prowess of prompt engineering becomes not just useful but essential. It acts as a crucial bridge between human intention and AI comprehension, ensuring interactions are productive, accurate, and goal-aligned.

The significance of prompt engineering is particularly evident in the realm of AI-generated art, where it underscores the indispensable role of human creativity. A notable example is the piece "Théâtre D'opéra Spatial," which clinched an award at a Colorado art competition. Crafted using Midjourney under the creative direction of co-creator Jason Allen, this artwork exemplifies the impact of ingeniously devised prompts. This instance illustrates how human ingenuity in prompt engineering can steer AI to create visually captivating art, affirming that despite AI's technological strides, the human element in crafting prompts remains pivotal in the artistic process.

Similarly, German artist Boris Eldagsen's triumph in the creative open category at the Sony World Photography Awards with an AI-generated image highlights the craft of prompt engineering in producing art on par with human-generated works. Eldagsen's success is a testament to the critical importance of skilled prompt crafting in digital artistry, where the creative visions of artists are key in directing AI to realize specific aesthetic visions.

A third case study features Anthropic's Claude 2.1, an AI model renowned for its expansive 200K token context window. This model exemplifies the transformative impact of prompt engineering on AI functionality. Through strategic prompting, users can navigate Claude 2.1's potential limitations, guiding it to process information more effectively. This case underscores the indispensable role of prompt engineering in enhancing AI's capabilities, highlighting that the quality of interaction is as crucial as the AI's technical specifications.

These case studies collectively underscore the growing importance of prompt engineering across various applications. They demonstrate that the ability to craft effective prompts is a distinct and vital skill, especially in areas where AI's creative and analytical capacities are harnessed to achieve outcomes that resonate with human expectations and artistic endeavors.

Prompt Engineering: A Universal Skill in the AI Era

In the ever-expanding domain of artificial intelligence, the practice of prompt engineering transcends the notion of a specialized profession, asserting itself as an essential skill for the modern workforce. This critical competency, which involves the nuanced craft of formulating effective instructions for generative models like ChatGPT, Bard, and others, is fast becoming as indispensable as traditional office tools like Microsoft Word or Excel.

From Niche to Necessity: The Evolution of Prompt Engineering

Prompt engineering is witnessing a pivotal shift from a niche expertise to a fundamental professional competency. Initially perceived as a domain reserved for specialists, the role of prompt engineers has become crucial in spearheading generative AI initiatives, serving as the cornerstone for integrating AI technologies across diverse sectors.

However, akin to the historical transition of computer literacy and typing skills from specialized abilities to baseline expectations, prompt engineering is on a similar trajectory towards becoming a universal professional requirement. This evolution mirrors the journey of skills such as social media proficiency or basic internet navigation, which have evolved from specialized or leisure activities to essential elements of contemporary personal and professional life.

Consider the integration of tools like Figma, Jira, and Git within specific professional domains; while indispensable for particular tasks, proficiency in these tools complements rather than defines professional roles. Similarly, the capability to engage effectively with generative AI models is emerging as a critical addition to the professional toolkit, transcending traditional domain boundaries.

The Democratization of Prompt Engineering

As we move deeper into the AI era, the democratization of prompt engineering marks a significant shift from a skill possessed by a select few to a universal competency, akin to computer literacy in today's professional landscape. The burgeoning demand for specialized prompt engineers is giving way to a future where the ability to interact with and leverage AI technologies is a widespread professional skill.

This democratization signifies a profound change in how we interact with technology, where the ability to craft effective AI prompts becomes a standard requirement. Just as the ability to navigate a word processor or spreadsheet is now considered fundamental, prompt engineering is set to become an indispensable skill, crucial for navigating the complexities and harnessing the opportunities of the AI-driven world.

The path forward for prompt engineering is unequivocal: evolving from a specialized discipline to a ubiquitous skill necessary for professional success. As AI continues to infiltrate every aspect of industry and commerce, the proficiency in prompt engineering will not only enhance our ability to communicate with AI but also unlock new dimensions of innovation and efficiency, marking a cornerstone of expertise in the dynamic and AI-enhanced future of work.

Mastering Prompt Engineering: A Pathway from Novice to Expert

In the book "Prompt Design Patterns", a detailed journey is meticulously outlined across three pivotal levels, offering a roadmap through the complex terrain of AI communication. This exploration delves into the Basic, Advanced, and Expert stages, revealing how our interactions with AI become increasingly refined at each level.

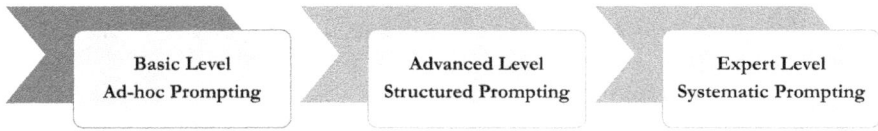

Figure 16-2: The Three Levels of Prompt Engineering

1. Basic Level — Laying the Foundations of AI Interaction

At the outset, prompt engineering resembles learning the fundamentals of a language — understanding its alphabets before moving on to words and sentences. This initial stage is marked by:

- **Simple Commands**: Engagements with AI are straightforward and task-specific. Users might request a weather forecast, translate a sentence, or tackle a basic arithmetic problem.

- **Direct Responses**: The AI's replies are literal and uncomplicated, mirroring the simplicity of the prompts.

- **Foundational Learning**: This stage is pivotal for grasping the AI's operational boundaries and capabilities, offering insights into how it perceives and reacts to elementary commands.

- **A Gateway for Novices**: Tailored for AI newcomers, this level demands no pre-existing knowledge of AI operations.

This foundational level is indispensable, establishing the bedrock for more elaborate AI interactions and paving the way for advanced communication strategies.

2. Advanced Level — Elevating AI Dialogues

Ascending to the advanced level, users adopt a more strategic lens for prompt formulation. This phase is characterized by:

- **Sophisticated Techniques**: Employing methods such as role-playing or context-setting to deepen AI interactions. Scenarios or narratives are crafted to steer AI responses.

- **Elevated Interaction**: Prompts are meticulously designed to draw out more detailed and context-sensitive replies from the AI.

- **Utilizing Templates and Tools**: The adoption of standardized frameworks and instruments aids in structuring prompts more cohesively, fostering consistent and reliable AI feedback.

- **Deciphering AI's Intricacies**: A profound comprehension of AI's language and contextual processing is essential. This stage demands an analytical and thoughtful approach to prompt crafting.

The advanced level unveils the expansive dialogue capabilities with AI, showcasing the profound interactions attainable between humans and machines.

3. Expert Level — Mastering the Art of AI Communication

The pinnacle of prompt engineering is the expert level, where the dialogue with AI evolves into sophisticated exchanges that transcend mere queries. Distinctive features of this stage include:

- **Complex Prompt Construction**: Here, prompts are akin to advanced programming, where users employ intricate design patterns to navigate complex AI dialogues.

- **Precision-Tuning AI Models**: Analogous to fine-tuning a musical instrument, this level involves the careful calibration of AI models to achieve peak performance.

- **Addressing AI Challenges**: Experts are attuned to potential issues such as AI-generated inaccuracies, biases, and privacy concerns, implementing robust validation, reliability enhancement, and debiasing strategies.

- **Innovative Prompt Techniques**: Methods like prompt chaining and versioning facilitate the creation of complex AI applications, with continuous evaluations to maintain their forefront position.

At the expert tier, prompt engineering goes beyond mere communication, entering domains where AI can deliver deep, human-like insights and reasoning.

As depicted in "Prompt Design Patterns", progressing through the levels of prompt engineering is both a formidable and enriching journey. From the essential building blocks to the nuanced complexities of expert interactions, each stage unveils new dimensions of AI's capabilities and potential. By mastering these tiers, we not only refine our communicative efficacy with AI but also unlock the vast array of its possibilities, transforming our engagements from basic commands to rich, nuanced dialogues.

Why Prompt Design Patterns is the Best Way to Master Prompt Engineering

Numerous methods exist to master prompt engineering, yet many fall short when it comes to meeting the specific and nuanced needs of diverse AI applications. Against this backdrop, "Prompt Design Patterns" stands out as a holistic solution, providing an in-depth exploration of both the art and science behind prompt engineering.

Limitations of Prompt Libraries

- **Generic Solutions**: Prompt libraries typically offer a broad range of prompts that, while useful, may not cater to the specific demands of unique or complex scenarios, leading to a compromise in the effectiveness of AI responses.

- **Limited Contextual Adaptability**: These prompts often lack the flexibility to adjust to the diverse contexts or the intricate nuances of various tasks, potentially resulting in less optimal AI interactions.

- **Dependency and Creative Stagnation**: An overreliance on pre-existing prompts can inhibit personal growth in prompt engineering skills, curtailing innovation and the development of bespoke prompts for novel situations.

Limitations of Prompt Tools

- **Technical Constraints**: AI-powered prompt generators may not fully grasp the intricate user requirements, occasionally missing the subtleties or the precise intentions behind a prompt.

- **Learning Curve**: Despite their intended user-friendly design, these tools can present a learning curve, necessitating users to familiarize themselves with their functionalities to craft effective prompts.

- **Opaque Operations**: Many prompt tools function as "black boxes," providing little to no insight into their operational mechanics, which can hinder users aiming to understand the foundational aspects of prompt generation.

The Unmatched Benefits of Prompt Design Patterns

- **Tailored and Flexible Prompt Crafting**: This approach empowers users with the methodologies to create highly customized prompts, adaptable across various scenarios and contexts, ensuring optimal effectiveness.

- **Foundational Skills and Deep Insights**: The book focuses on building a solid foundation in prompt engineering principles, enabling users to understand and construct effective prompts from scratch, fostering skill development.

- **Encouragement of Innovation**: Through the application of design patterns, users are inspired to employ creativity and innovation, essential for navigating the dynamic AI landscape.

- **Strategic Approach to Complex Challenges**: Offering strategic frameworks, "Prompt Design Patterns" equips users to tackle sophisticated problems in AI interactions, surpassing the basic solutions offered by other resources.

- **Sustainable Mastery and Future-Proof Skills**: The knowledge and skills acquired from this book are designed to endure, ensuring users remain proficient and relevant as AI technology advances.

- **Bridging the Knowledge Gap**: By addressing the gaps left by other tools, the book grants users comprehensive control over the prompt crafting process, leading to superior AI engagements.

- **Enhanced Collaboration and Knowledge Sharing**: The standardized design patterns facilitate effective collaboration among professionals, streamlining communication and improving collective productivity.

- **Forward-Thinking and Strategic Innovation**: Beyond addressing current challenges, the book anticipates future developments in AI, preparing users to innovate proactively.

"Prompt Design Patterns" is not just a compilation of theories but a consolidation of 23 research-backed design patterns, each borne out of extensive expertise and documented successes in AI research. These patterns represent the culmination of best practices in prompt engineering, providing a practical, structured, and insightful guide to mastering AI dialogues. From basic communication strategies to advanced techniques that redefine the possibilities of generative AI, the book is a beacon for those aspiring to excel in the intricate ballet of human-AI interaction.

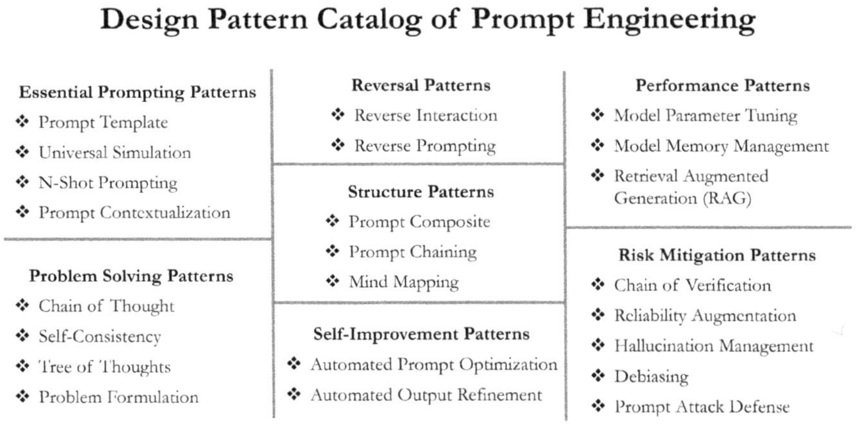

Design Pattern Catalog of Prompt Engineering

Essential Prompting Patterns	Reversal Patterns	Performance Patterns
❖ Prompt Template	❖ Reverse Interaction	❖ Model Parameter Tuning
❖ Universal Simulation	❖ Reverse Prompting	❖ Model Memory Management
❖ N-Shot Prompting	**Structure Patterns**	❖ Retrieval Augmented Generation (RAG)
❖ Prompt Contextualization	❖ Prompt Composite	
	❖ Prompt Chaining	**Risk Mitigation Patterns**
Problem Solving Patterns	❖ Mind Mapping	❖ Chain of Verification
❖ Chain of Thought		❖ Reliability Augmentation
❖ Self-Consistency	**Self-Improvement Patterns**	❖ Hallucination Management
❖ Tree of Thoughts	❖ Automated Prompt Optimization	❖ Debiasing
❖ Problem Formulation	❖ Automated Output Refinement	❖ Prompt Attack Defense

Figure 16-3: The 23 Prompt Design Patterns from the book "Prompt Design Patterns"

For enthusiasts and professionals alike, "Prompt Design Patterns" is the definitive resource for mastering AI communication, offering a blend of insights, strategies, and real-world applications to navigate the complexities of prompt engineering.

Integrating Domain Knowledge with AI Prompt Engineering for the Next-Generation Workforce

Prompt engineering distinguishes itself through its reliance on domain knowledge. For instance, when creating prompts for a travel itinerary, the richness and relevance of the responses significantly depend on the user's understanding of travel logistics, destinations, and preferences. Without such knowledge, the prompts might not elicit detailed and useful information.

Professionals must also recognize the limitations of Large Language Models (LLMs) like ChatGPT, which can sometimes produce misleading or incomplete answers. It's the deep domain expertise that enables a user to critically evaluate the accuracy and completeness of an AI model's outputs.

The integration of domain knowledge with prompt engineering in AI mirrors the scenario of a master chef utilizing a state-of-the-art kitchen. Just as a chef applies their extensive culinary skills with advanced kitchen technologies to create exceptional dishes, professionals across various fields can similarly merge their specialized knowledge with prompt engineering to enhance AI's capabilities.

Consider a chef specializing in Italian cuisine, knowledgeable about the subtleties of Italian cooking techniques and ingredients. If this chef has access to an AI-powered kitchen but lacks the skills to interact with advanced culinary tools, the kitchen's potential is not fully realized. Conversely, an operator proficient in using the AI kitchen but without culinary expertise may produce dishes that, while technically sound, lack authenticity and finesse.

Domain Knowledge and Prompt Engineering: A Synergistic Relationship

In the medical field, a physician's understanding of medical conditions, treatments, and terminology, combined with prompt engineering skills, enables effective communication with AI systems for analyzing patient data, suggesting diagnoses, or drafting patient information. This ensures AI-generated responses are not just accurate but also contextually relevant and practically applicable.

In software development, domain knowledge encompasses programming languages, algorithms, and project-specific challenges. Prompt engineering allows developers to guide AI tools in code generation, debugging, and design, enhancing efficiency and creativity without supplanting the developer's expertise.

For educators, deep pedagogical and subject matter knowledge, paired with prompt engineering, facilitates the creation of bespoke learning materials or interactive educational experiences. This approach ensures AI-generated content meets specific learning goals, student needs, and educational standards.

The Harmonious Fusion of Human Expertise and AI Capabilities

The convergence of domain knowledge and prompt engineering transcends mere collaboration with AI; it's about forging a seamless integration where human expertise steers AI towards outcomes that are meaningful, accurate, and contextually apt. This partnership is akin to a symphony orchestra, where the conductor's direction brings together each musician's talent to produce a masterpiece greater than the sum of its parts. Through this synergy, the next-generation workforce can unlock the full potential of AI, leveraging it to achieve unprecedented levels of innovation and efficiency.

Envisioning the Future of Human-AI Synergy

As we stand at the precipice of a transformative era in communication, the concept of centaur chess, blending human intellect with AI's computational might, becomes a beacon for the future of human-AI collaboration. This chapter has journeyed through the intricate dynamics of this partnership, underscoring the pivotal role of prompt engineering as the cornerstone of initiating the AI transformation journey.

The dawn of this new communication era invites us to reimagine the possibilities of interaction, where the centaur model serves not just as a strategy but as a vision for enhancing the depth and breadth of our communicative efforts. It represents a harmonious blend of human emotional intelligence and AI's vast data-processing capabilities, elevating the efficiency and impact of our exchanges. In this landscape, prompt engineering stands as the critical conduit, enabling a seamless flow between human intuition and AI's analytical precision.

This fusion, facilitated by masterful prompt crafting, empowers us to guide AI towards generating responses that are both contextually relevant and emotionally resonant. It transforms AI from a mere tool into a nuanced partner in communication, capable of adapting to the diverse needs and preferences of users. Through this lens, we see not a replacement of human capabilities but an enhancement, where AI complements and amplifies the richness of human interaction.

Looking forward, the integration of human capabilities and AI holds the promise of revolutionizing how we connect, share, and understand each other. The adoption of the centaur model in communication strategies heralds a future where AI is embraced as an ally, enriching human experiences and interactions. This paradigm shift, grounded in the principles of prompt engineering, invites us to embark on an AI transformation journey, not as competitors but as collaborators, aiming to create a more connected, empathetic, and intelligent world.

In this era of rapid technological advancement, prompt engineering emerges not just as a skill but as a foundational step towards harnessing the full potential of AI in communication. It beckons us to start our AI transformation journey with a focus on enhancing the synergy between human creativity and AI efficiency, paving the way for a future where our communicative endeavors are not just enhanced but transformed.

Chapter 17

Transforming to an AI Native Enterprise

"Change is inevitable, but transformation is a choice." — *Heather Ash Amara*

I n an era where technological advancement races ahead at an unprecedented pace, the quest to become an AI Native Enterprise stands as the definitive journey for organizations aiming to not only survive but thrive. As we delve into the last chapter of this insightful book, we confront the crucial intersection of innovation and transformation, a nexus at which businesses must navigate the complex, ever-evolving landscape of Artificial Intelligence (AI).

Why do some organizations succeed in seamlessly integrating AI into their fabric, transforming operations, customer experiences, and industry standards, while others stumble or falter? What distinguishes a business that leverages AI to redefine its future from one that merely adopts technology as a peripheral tool? These questions are not just theoretical; they are deeply practical, addressing the heart of the challenges and opportunities presented by AI today.

The journey towards becoming an AI Native Enterprise is fraught with dilemmas and decisions. It requires a strategic overhaul not just of technology infrastructure, but of organizational culture, business processes, and market positioning. The transformation is as much about adopting new technologies as it is about fostering a mindset of continuous innovation, agility, and resilience.

In this final chapter, we synthesize the essence of our exploration into AI and its transformative impact on businesses across sectors. We present a comprehensive guide for navigating the AI transformation journey, drawing on a rich tapestry of frameworks and approaches that have illuminated our path so far. From the Dual Transformation model, which emphasizes balancing core optimization with innovation, to the Five Frames of Performance and Health, offering a step-by-step approach to managing change, and the strategic pathways delineated in the Business-Driven AI-Native Transformation Framework, we aim to equip you with the knowledge and tools to embark on this transformative journey.

Whether you are leading a small startup keen on leveraging AI for growth, a medium-sized enterprise exploring new markets and opportunities, or a large conglomerate aiming to maintain competitive advantage and industry leadership, this chapter offers insights tailored to your unique journey towards AI integration.

As we traverse the landscape of AI transformation, remember that the path is not linear nor the journey solitary. It is a shared voyage of discovery, innovation, and reinvention. Let us embark on this final chapter not just as a conclusion to our book, but as the beginning of a new chapter in your organization's story — a story of transformation, resilience, and success in the AI-driven future.

Understanding AI Native Transformation

An AI Native Enterprise is characterized by its intrinsic ability to embed AI across its operations, decision-making processes, and product offerings. This transformation goes beyond mere technological adoption; it entails a fundamental rethinking of business models, workflows, and corporate culture to harness the full potential of AI.

Understanding AI Native Transformation involves comprehensively grasping how the integration and strategic use of Artificial Intelligence (AI) can fundamentally change an enterprise's operations, value proposition, and competitive stance. An AI Native Enterprise is not merely one that utilizes AI tools or incorporates AI in a handful of processes; it is an organization that embeds AI at the core of its business strategy, operational processes, and culture. This transformation impacts every facet of the organization, from how decisions are made to how products and services are designed and delivered.

Core Components of AI Native Transformation

- **Strategic Integration of AI**: AI Native Enterprises strategically deploy AI technologies to enhance their competitive advantage, not just for incremental improvements but as a transformative force across the business landscape. This involves leveraging AI to unlock new business models, enter new markets, and create differentiated value for customers.

- **Operational Excellence through AI**: At the operational level, AI technologies are embedded into the fabric of business processes. This could mean automating routine tasks with AI to improve efficiency, using AI to optimize supply chain logistics, or deploying AI-driven analytics to forecast market trends and inform strategic decisions. The goal is to achieve a higher level of operational efficiency, agility, and responsiveness to market dynamics.

- **Data-Driven Culture**: An AI Native Enterprise fosters a culture that values data as a critical asset. This involves cultivating data literacy across the organization and ensuring that decisions at all levels are informed by data and AI-generated insights. A data-driven culture empowers employees, enhances collaboration, and facilitates innovation.

- **AI as a Product and Service Innovator**: Beyond internal efficiencies, AI Native Enterprises leverage AI to innovate their product offerings and how they engage with customers. This could involve developing AI-powered products or incorporating AI into existing products to enhance their functionality and appeal. It also means using AI to personalize customer interactions and improve the customer experience through recommendations, predictive services, and automated support.

- **Ethical and Responsible Use of AI**: Understanding AI Native Transformation also involves recognizing the importance of ethical considerations. AI Native Enterprises commit to using AI responsibly, ensuring that AI systems are transparent, fair, and respect privacy. This ethical approach is crucial for building trust with customers, employees, and stakeholders.

- **Continuous Learning and Adaptation**: Finally, becoming an AI Native Enterprise is an ongoing journey of learning and adaptation. It requires continuous investment in AI research and development, staying abreast of emerging AI technologies, and adapting business strategies in response to new AI-driven opportunities and challenges.

The Transformational Impact

The transformation to an AI Native Enterprise has far-reaching implications:

- **Competitive Advantage**: Organizations that effectively harness AI can outperform competitors by being more innovative, efficient, and responsive to customer needs.

- **Innovation and Growth**: AI opens up new avenues for growth through innovative products, services, and business models that were previously unimaginable.

- **Customer Engagement**: AI enables a deeper understanding of customer preferences and behaviors, leading to enhanced customer experiences and loyalty.

- **Operational Efficiency**: AI-driven automation and optimization can lead to significant cost savings and operational improvements, allowing organizations to allocate resources more strategically.

Understanding AI Native Transformation is about recognizing the strategic importance of AI in shaping the future of business. It's a comprehensive approach that goes beyond technology adoption, embedding AI into the very DNA of an organization to drive sustainable growth and innovation.

The Business-Driven AI-Native Transformation Framework

Chapter 9 introduces the AI-Powered Business Transformation Playbook, a strategic guide delineating the evolutionary journey of enterprises through artificial intelligence integration to redefine their operations, business models, and industry standings. This playbook, rooted in business strategy, unfolds across three transformative stages, guiding enterprises from enhancing current operations with AI to fundamentally reinventing their business landscapes.

Stage 1: Business Augmentation
This foundational stage centers on integrating AI to enhance and optimize existing business processes. The focus is on leveraging AI for automating routine tasks and employing predictive analytics for improved decision-making. Objectives include streamlining operations, boosting customer experiences, and increasing overall efficiency. Examples of augmentation include AI-powered chatbots for customer service and predictive analytics for supply chain optimization, laying the groundwork for deeper AI integration.

Stage 2: Business Transformation
Upon establishing foundational AI capabilities, enterprises embark on transforming their business models. This stage involves embedding AI into core operations to innovate products, personalize customer experiences, and enter new markets. AI-driven insights enable ventures into adjacent spaces and the creation of platform ecosystems, exemplified by AI-driven retail platforms with virtual try-on features

and smart healthcare solutions offering personalized care through telehealth services.

Stage 3: Business Reinvention
The pinnacle of the AI transformation journey empowers companies to use generative AI for creating new business models and industries. This stage is about envisioning and incubating disruptive innovations to redefine the enterprise and its offerings. Ventures might include AI-driven urban planning for smarter cities or generating interactive, adaptive content in the entertainment industry, pioneering future possibilities with generative AI.

While these stages can be approached sequentially, with each building upon the last to expand AI's scope and impact, organizations might also choose an ambitious route to tackle all three simultaneously, based on various factors.

In Chapter 8, we outlined the AI Adoption Roadmap, a structured guide through five phases of integrating AI into organizational culture and operations:

- **Planning Phase**: Identifying AI use cases, developing infrastructure, securing executive support, and building an AI-knowledgeable team.

- **Pilot Phase**: Testing selected use cases through prototypes, setting up AI infrastructure, and refining strategies based on iterative learning cycles.

- **Institutionalization Phase**: Operationalizing AI with a focus on governance, establishing a Center of Excellence, and embedding AI capabilities for responsible scaling.

- **Scaling Phase**: Expanding AI application enterprise-wide, enhancing infrastructure, and advancing automation for efficiency.

- **Differentiation Phase**: Integrating AI into strategic decision-making, products, and services for competitive advantage, marked by enhanced revenue and market share.

Integrating the AI Adoption Roadmap with the AI-Powered Business Transformation playbook creates a comprehensive and adaptive framework for a business-driven, AI-native transformation. This holistic approach covers the full spectrum of business and technology challenges, leveraging a hybrid model to maximize AI's potential. It ensures AI adoption is strategic, aligned with organizational maturity, and conducive to sustainable value realization, setting the foundation for enterprises to thrive as AI-native entities in a rapidly evolving market landscape.

Harmonizing AI Adoption and Business Transformation: Pathways to an AI-Native Future

Integrating the AI Adoption Roadmap with the AI-Powered Business Transformation playbook involves synchronizing the technological adoption process with strategic business transformation goals. This integration can be approached in multiple ways, each offering unique advantages to organizations aiming to become AI-native entities. Here are several methods for integration, along with their unique advantages:

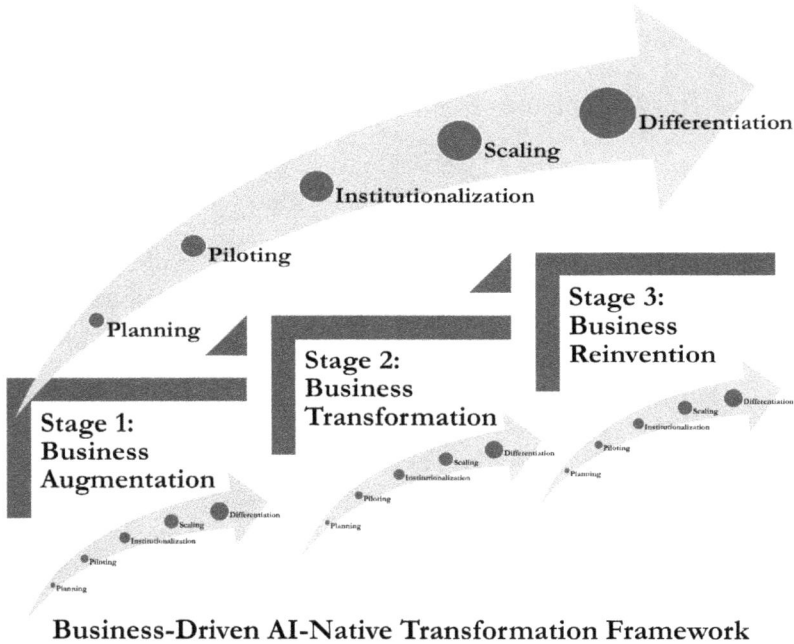

Business-Driven AI-Native Transformation Framework

Figure 17-1: The Business-Driven AI-Native Transformation Framework

- **Sequential Integration** focuses on organizations progressing through the AI Adoption Roadmap phases—Planning, Pilot, Institutionalization, Scaling, Differentiation—before embarking on the stages of business transformation: Augmentation, Transformation, Reinvention. This method lays a solid AI foundation across the organization, minimizing risks associated with transformative AI applications and allowing for measured progress through clear milestones and easier transformation management.

- **Parallel Integration** enables organizations to simultaneously navigate the AI Adoption Roadmap and implement business transformation strategies. By aligning AI projects with strategic business objectives from the start, this approach not only accelerates the transformation into an AI-native entity but also ensures immediate value realization from AI investments, enhancing organizational support for the transformation.

- **Modular Integration** adopts a flexible, unit-specific approach, where different business units or functions progress through the AI Adoption Roadmap based on their unique transformation goals. This strategy acknowledges the varied needs and readiness levels within the organization, offering tailored AI implementation strategies and mitigating risks by allowing organizational segments to advance at their own pace.

- **Ecosystem-Driven Integration** expands the transformation effort beyond the organization to include partners, suppliers, and customers. By aligning AI adoption and business transformation stages with ecosystem development, this method not only broadens innovation potential but also facilitates shared learning and risk management, creating a collaborative approach to overcoming transformation challenges.

Regardless of the chosen method, the integration of the AI Adoption Roadmap with the AI-Powered Business Transformation playbook enables organizations to approach AI as both a technological and strategic asset. This comprehensive framework supports sustainable, business-driven AI-native transformation, ensuring that AI investments are directly linked to strategic business outcomes, operational efficiency, and market differentiation. The unique advantage of any integration approach is its ability to create a structured yet flexible pathway to becoming an AI-native enterprise, capable of continuously innovating and adapting in an AI-driven market landscape.

While our newly introduced Business-Driven AI-Native Transformation Framework significantly boosts the likelihood of successful AI transformation within organizations, orchestrating large-scale change remains a formidable challenge. Industrial research consistently shows that approximately 70% of transformation initiatives do not achieve their intended outcomes. To enhance the prospects of success in your AI transformation journey, the following sections will unveil three additional research-supported frameworks. These frameworks offer diverse strategies for undertaking AI transformation, each providing unique insights and methodologies to navigate the complexities of change.

AI-Native Transformation: The Dual Transformation Approach

For organizations navigating the complex journey of AI integration, the Dual Transformation framework, originally conceptualized by Scott D. Anthony, Clark G. Gilbert, and Mark W. Johnson in their work on organizational change and innovation, offers a strategic blueprint. This approach outlines how enterprises can evolve into AI Native Enterprises, using AI not just for automation but as a foundational element for strategic innovation and operational excellence.

This section will explore the pathway to becoming an AI Native Enterprise through the Dual Transformation framework. This strategy advocates for two concurrent paths: the first strengthens and refines existing business operations with AI, aiming for efficiency and resilience; the second propels the organization into new realms of innovation and market opportunities powered by AI. By pursuing these parallel paths, businesses can fully leverage the transformative power of AI, moving beyond adoption to become frontrunners in the AI-driven future.

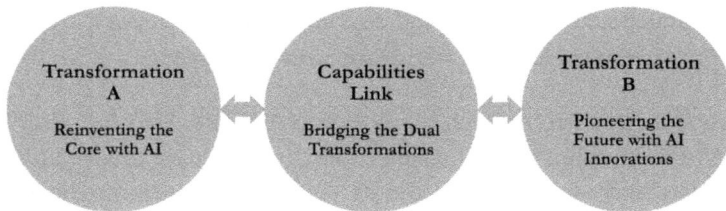

Figure 17-2: The Dual Transformation Approach for AI Transformation

Transformation A: Reinventing the Core with AI

At the heart of becoming an AI Native Enterprise lies the pivotal process of reinvigorating the core business through the strategic integration of Artificial Intelligence (AI). This foundational transformation, known as Transformation A, aims to infuse existing business frameworks, processes, and products with AI, thereby enhancing operational efficiency, customer satisfaction, and decision-making capabilities. It's a deliberate strategy designed to ensure the core business not only remains competitive but also resilient amid the fast-paced technological advancements and market shifts.

Key Strategies for Transformation A

- **Automate for Efficiency**: By automating routine and repetitive tasks, businesses can streamline operations, reduce costs, and allocate human talent to more strategic initiatives. From data entry to customer service via AI-powered chatbots and predictive maintenance, automation lays the groundwork for a more efficient operational model.

- **Enhance Decision Making**: Leveraging machine learning models and advanced analytics, AI can process and extract actionable insights from vast datasets, thereby improving the accuracy and speed of business decisions. This strategic application of AI makes businesses more agile and better positioned to respond to market dynamics.

- **Personalize Customer Experiences**: AI's ability to analyze detailed customer data enables businesses to tailor experiences, products, and services to individual customer preferences, significantly enhancing customer engagement and loyalty.

- **Optimize Supply Chain and Operations**: Through predictive analytics, AI optimizes supply chain and operational efficiencies, forecasting demand, managing inventory levels, and improving delivery routes. Additionally, AI enhances operational productivity by identifying and mitigating potential bottlenecks and equipment failures.

- **Cultivate an AI-Enabled Workforce**: Essential to Transformation A is the development of an AI-literate workforce capable of leveraging AI insights, managing AI systems, and integrating AI tools within their daily tasks. This upskilling fosters an adaptable and innovative organizational culture.

Implementing Transformation A

To implement Transformation A effectively, organizations must:

- **Identify High-Impact Areas**: Focus initially on processes where AI can significantly enhance efficiency, customer experience, or revenue generation.

- **Build Robust Data Infrastructure**: A solid data infrastructure is crucial for supporting AI initiatives, necessitating comprehensive data collection, storage, and analysis capabilities.

- **Foster Innovation Culture**: Cultivating a culture that embraces experimentation and lifelong learning is vital for exploring and integrating AI across the business.

- **Address Ethical Considerations**: Ethical and social implications of AI, including transparency, fairness, and privacy, must be diligently considered to ensure responsible AI use.

- **Monitor and Iterate**: Continuous monitoring and iterative refinement of AI initiatives are essential for leveraging insights and enhancing performance over time.

Transformation A transcends mere technology adoption, requiring a strategic reevaluation of core business processes through an AI lens. This shift towards operational excellence and innovation ensures that businesses not only maintain their competitive edge but also lay a solid foundation for sustained growth and transformation in the AI era.

Transformation B: Pioneering the Future with AI Innovations

As enterprises solidify their current operations through AI in Transformation A, Transformation B propels them toward the horizon, leveraging AI to forge new paths in market exploration, product innovation, and business model reinvention. This phase is about transcending the incremental improvements offered by AI to existing processes, and instead, embarking on a venture into entirely uncharted territories of business opportunities that AI technologies unlock.

Key Strategies for Transformation B

- **Innovate Products and Services**: AI's capability to sift through and analyze trends, behaviors, and gaps in customer needs presents unparalleled opportunities for innovation. Enterprises can harness AI to develop intelligent products that intuitively adapt to user preferences or predictive services that anticipate customer needs, creating personalized health platforms or AI-driven advisory services as prime examples.

- **Explore New Markets**: With AI's analytical prowess, businesses can identify untapped or underserved markets, predicting emerging demands for products and services. This insight provides a strategic edge, enabling entry into new markets with tailored offerings that meet unique local needs.

- **Redefine Business Models**: The transformative power of AI allows businesses to overhaul their traditional business models. Companies might transition from selling products to offering AI-enhanced services on a subscription basis, continually adapting and personalizing based on AI-derived insights.

- **Foster Collaborative Ecosystems**: AI paves the way for building ecosystems that unite partners, suppliers, and even competitors in a collaborative innovation network. Within these ecosystems, AI optimizes interactions and operations, such as dynamically adjusting supply chains to meet real-time changes in demand or conditions.

- **Cultivate a Culture of Innovation**: The leap into the future with AI demands a corporate culture steeped in experimentation, risk-taking, and perpetual learning. Encouraging teams to cross-collaborate on AI projects, investing in AI research, and creating an environment that celebrates innovation are pivotal for nurturing Transformation B.

Implementing Transformation B

For a successful transition into Transformation B, enterprises should:

- **Invest in AI Research and Development**: Dedicate resources to uncover and experiment with emerging AI technologies that hold the potential to open new business avenues.

- **Prototype and Test Rapidly**: Embrace agile methodologies for swift prototyping, testing, and refining of AI innovations, ensuring a cycle of continuous learning and improvement.

- **Engage with the Innovation Ecosystem**: Forge partnerships with startups and academic circles to remain at the cutting edge of AI innovation, tapping into a wellspring of fresh ideas and talent.

- **Navigate Ethical and Regulatory Landscapes**: As enterprises venture into novel AI applications, it's crucial to thoughtfully address regulatory and ethical considerations, ensuring innovations are sustainable and responsible.

- **Prepare for Organizational Evolution**: The introduction of new AI-driven business areas may necessitate structural, processual, and skill-based changes within the organization, requiring thorough preparation and adaptation strategies.

Transformation B envisions a future where AI is not just a support mechanism but the main driver of business growth and innovation. It challenges enterprises to boldly reimagine the possibilities, invest in breakthrough ideas, and pivot as needed to seize new market opportunities and customer needs. By embracing Transformation B, AI Native Enterprises unlock unparalleled potential, securing their growth and relevance in an increasingly AI-centric world.

The Capabilities Link: Bridging the Dual Transformations

At the core of transitioning into an AI Native Enterprise lies the Capabilities Link, a pivotal element within the Dual Transformation framework. This strategic bridge seamlessly connects Transformation A, which focuses on revitalizing the core business through AI, and Transformation B, dedicated to pioneering future innovations with AI. It comprises shared assets, expertise, and insights that fuel both transformations, fostering synergy, enhancing efficiency, and accelerating innovation. This interconnection ensures that efforts to strengthen the core business and to venture into new opportunities are not disparate strategies but integral parts of a unified approach towards becoming an AI Native Enterprise.

Components of the Capabilities Link

- **Shared Data and Analytics**: At the foundation, shared data infrastructure and analytics capabilities play a vital role. Data derived from everyday business operations powers AI innovations, while insights from AI projects enhance existing processes. This symbiotic exchange of data and insights underpins a data-driven transformation, bolstering strategic decision-making.

- **Technology Platforms**: AI Native Enterprises thrive on flexible and scalable technology platforms that support both the optimization of core processes and the development of new applications. These platforms, encompassing cloud computing, AI frameworks, and development tools, enable swift innovation and minimize redundant efforts across the enterprise.

- **Talent and Expertise**: The human factor—encompassing skills, knowledge, and the organizational culture—is critical. Cultivating a workforce proficient in AI and data science and a culture that prizes innovation and lifelong learning is essential for both Transformation A and B. This dual focus requires hiring and nurturing technical talent while also instilling a mindset of flexibility and innovation organization-wide.

- **Innovation Processes and Governance**: Robust innovation management processes and agile governance structures are essential for linking the two transformations. They allow for the rapid testing and scaling of ideas, ensure appropriate risk management, and align investments with strategic objectives. Shared governance models balance the operational needs of the core business with the exploratory demands of new initiatives.

Implementing the Capabilities Link

For the Capabilities Link to effectively fulfill its role, enterprises should:

- **Encourage Cross-Functional Collaboration**: Promote a culture where teams engaged in core business improvements and those exploring AI innovations share insights and learnings, fostering unexpected synergies.

- **Invest in Scalable Technologies**: Embrace technologies that can be easily scaled or adapted, ensuring that investments support both existing operations and future innovations.

- **Develop a Learning Organization**: Establish ongoing education and skill development programs, enabling employees to contribute to both facets of the transformation.

- **Balance Exploration and Exploitation**: While focusing on refining the core business, also dedicate resources to exploring new opportunities, using the Capabilities Link to facilitate the effective flow of resources and insights between these areas.

The Capabilities Link transcends being merely a collection of shared resources; it represents a strategic methodology to ensure that the enhancement of the core business through AI and the exploration of new AI-driven opportunities mutually reinforce each other. By leveraging this link adeptly, AI Native Enterprises can strike a balance between optimizing their current operations and fostering innovation for the future. This equilibrium ensures enduring growth and a competitive edge in the swiftly evolving digital domain.

Conclusion

The journey toward becoming an AI Native Enterprise, as delineated through the Dual Transformation framework, outlines a visionary yet pragmatic path for organizations navigating the complexities of the digital age. This approach, with its dual focus on revitalizing core operations (Transformation A) and pioneering innovative ventures (Transformation B), offers a holistic strategy for businesses aiming to harness the full potential of Artificial Intelligence.

Dual Transformation underscores the necessity of balancing operational excellence with strategic innovation, highlighting the importance of the Capabilities Link as the conduit that binds these two transformative processes. By fostering a symbiotic relationship between enhancing the core business and exploring new frontiers, organizations can ensure that their AI integration efforts are not only comprehensive but also synergistic, driving sustainable growth and competitive advantage.

AI-Native Transformation: The Dual Operating System Approach

Inspired by John Kotter's influential Dual Operating System concept as presented in "Accelerate", this section explores the third strategic pathway for organizations aiming to evolve into AI Native entities. This approach skillfully balances the pursuit of operational efficiency with the drive for innovation and agility, essential components in the age of Artificial Intelligence.

The Dual Operating System framework advocates for a bifocal operational mode: one part is the hierarchical system, ensuring the organization's efficiency and stability, and the other is the network system, fostering innovation and agility. This framework is particularly applicable to AI transformation, enabling businesses to seamlessly incorporate AI technologies into their existing operations while also venturing into new, AI-enabled innovative projects.

By adopting this dual approach, organizations can maintain their core business processes without sacrificing the flexibility needed to explore groundbreaking AI applications. This balance is crucial for staying competitive in a rapidly evolving technological landscape, ensuring that enterprises not only keep pace with current advancements but also shape future trends through innovation.

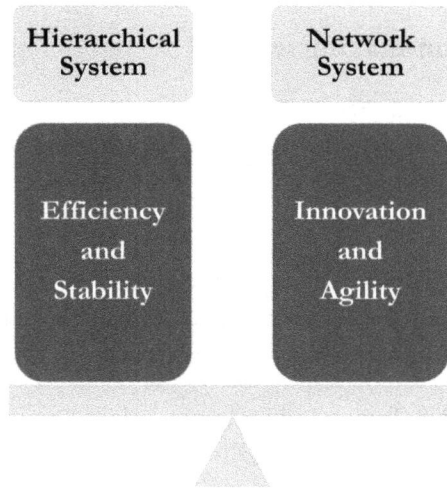

Figure 17-3: The Dual Operating System Approach for AI Transformation

Hierarchical System: Integration AI into Core Operations

The Hierarchical System within the context of transforming to an AI Native Enterprise involves embedding Artificial Intelligence (AI) technologies into the core operations of an organization. This integration aims to enhance efficiency, improve decision-making processes, and elevate customer engagement by leveraging AI's capabilities to automate, analyze, and personalize at scale. The hierarchical system, traditionally characterized by its structured and stable nature, serves as the backbone of an organization, ensuring that daily operations run smoothly and efficiently.

Automating Routine Processes for Enhanced Efficiency

- **Operational Efficiency**: AI's ability to automate repetitive tasks across departments—from finance and HR to customer service—eliminates time-consuming processes, thus freeing up valuable human resources for strategic initiatives.

- **Error Reduction**: Automation minimizes human error, ensuring operations are executed with precision and reliability, enhancing the quality of work and reducing the risk of costly mistakes.

- **Cost Savings**: By optimizing resource allocation and reducing the dependency on manual labor, automation directly contributes to significant cost reductions.

Enhancing Decision-making with AI

- **Data-driven Insights**: Utilizing machine learning and predictive analytics, AI sifts through and analyzes large datasets to reveal insights that drive strategic decisions, offering a deeper understanding of market trends and consumer behavior.

- **Real-time Analysis**: The capability for real-time data analysis allows for swift, informed decision-making, a critical advantage in the fast-paced business world where timing can be everything.

- **Customized Solutions**: AI systems tailored to the business's unique requirements offer customized insights, ensuring solutions are directly aligned with specific organizational challenges and opportunities.

Elevating Customer Engagement

- **Personalization at Scale**: Through detailed analysis of customer data, AI enables highly personalized experiences, from product recommendations to marketing communications, significantly enhancing customer satisfaction and loyalty.

- **Automated Customer Support**: By deploying AI-powered chatbots and virtual assistants, organizations can offer round-the-clock customer support, efficiently resolving inquiries and issues, thereby optimizing customer service operations.

- **Predictive Customer Behavior Analysis**: AI's predictive capabilities allow businesses to anticipate customer needs and preferences, facilitating proactive service and product offerings.

Integrating AI within an organization's hierarchical system is a strategic endeavor that strengthens the foundation for ongoing competitiveness and growth. This requires careful planning, robust technological infrastructure, and a culture of agility to navigate potential challenges and maximize the advantages of AI integration. Successfully embedding AI into core operations not only optimizes existing processes but also sets the stage for sustained innovation and market leadership in the digital era.

Network System: Catalyzing Innovation with AI

The Network System in the transformation to an AI Native Enterprise represents a dynamic, flexible, and innovative approach to exploring new AI-driven opportunities. Unlike the Hierarchical System, which focuses on embedding AI into existing processes for efficiency and stability, the Network System leverages AI to foster innovation, agility, and growth. It operates parallel to the traditional structure, forming a complementary force that drives the organization towards new frontiers and opportunities. This system is characterized by its ability to rapidly respond to market changes, explore new business models, and develop groundbreaking products and services powered by AI.

Innovating Products and Services with AI

- **AI-Powered Solutions**: The Network System spurs the creation of innovative products and services by harnessing AI capabilities like predictive analytics, natural language processing, and computer vision. This empowers businesses to address evolving customer needs and penetrate new market segments.

- **Customer-Centric Innovations**: Leveraging AI to analyze customer data and feedback in real-time, organizations can craft offerings that are highly personalized and tailored to individual preferences, enhancing customer engagement and satisfaction.

- **Rapid Prototyping and Testing**: Embracing an agile methodology, the Network System enables swift prototyping and testing of AI-driven innovations, significantly accelerating the journey from concept to market.

Pioneering New Business Models

- **AI-as-a-Service (AIaaS)**: This model allows organizations to offer AI capabilities as a service, opening new revenue channels and making advanced AI technologies accessible to a broader audience, including small businesses.

- **Platform-Based Ecosystems**: Utilizing AI, enterprises can establish interactive, platform-based ecosystems that connect users, providers, and stakeholders, offering personalized experiences and services that distinguish them in the marketplace.

- **Subscription and Usage-Based Models**: AI facilitates the adoption of flexible pricing models, such as subscriptions or usage-based billing, aligning more closely with customer usage patterns and preferences.

Cultivating Collaborative Ecosystems

- **Partnerships with Startups and Tech Firms**: Collaborating with startups and tech companies grants access to the forefront of AI research and innovation, fostering the development of unique AI applications.

- **Engagement with Academia**: Working alongside academic and research institutions connects businesses with the latest AI advancements and a pool of talent, accelerating the translation of AI research into practical applications.

- **Open Innovation Platforms**: Participation in open innovation platforms invites organizations to crowdsource ideas and solutions from a worldwide network of innovators, expanding the horizons of what's possible with AI.

The Network System plays a vital role in equipping organizations to explore and capitalize on AI-driven opportunities, ensuring they can innovate and grow. In conjunction with the Hierarchical System, it guarantees that enterprises can refine their current operations while simultaneously setting the pace for innovation in products, services, and business models. This dual strategy enables organizations to fully embrace their AI Native Enterprise potential, leveraging AI to navigate the future and secure success in the digital age.

Conclusion

The Dual Operating System Approach, inspired by John Kotter's groundbreaking concepts, represents a strategic blueprint for organizations embarking on the transformational journey to AI Native status. This method, blending the efficiency and stability of the Hierarchical System with the innovation and agility of the Network System, offers a comprehensive model for integrating AI across all facets of an enterprise. It encapsulates the essence of navigating modern business challenges—balancing the need for operational excellence with the imperative to innovate and adapt in an AI-driven era.

This approach underscores the importance of fostering a dynamic ecosystem within the organization—one that promotes cross-functional collaboration, leverages shared data and analytics, and cultivates a workforce skilled in navigating AI technologies. It's a call to action for leaders to champion a dual-focused strategy, ensuring that AI initiatives are deeply integrated into the organizational fabric, driving growth, and securing a competitive edge.

In navigating the path to becoming an AI Native Enterprise, the Dual Operating System Approach offers a balanced and flexible framework that acknowledges the complexities of modern business. It provides a roadmap for organizations to not only survive the waves of digital disruption but to ride them confidently into the future. By embracing this approach, enterprises can unlock new realms of possibility, fostering an environment where innovation flourishes and AI becomes a core driver of strategic success.

AI-Native Transformation: The Five Frames of Performance and Health Approach

Embarking on the path to becoming an AI Native Enterprise demands a strategic overhaul, integrating Artificial Intelligence (AI) seamlessly into both the operational and cultural fabric of an organization. Inspired by the insights from "Beyond Performance 2.0" by Scott Keller and Bill Schaninger, this section introduces the Five Frames of Performance and Health as the fourth comprehensive framework for guiding leaders through the intricacies of AI transformation. This methodology not only aims at achieving exceptional performance through AI but also emphasizes fostering a vibrant organizational culture that nurtures continuous innovation and sustainable growth.

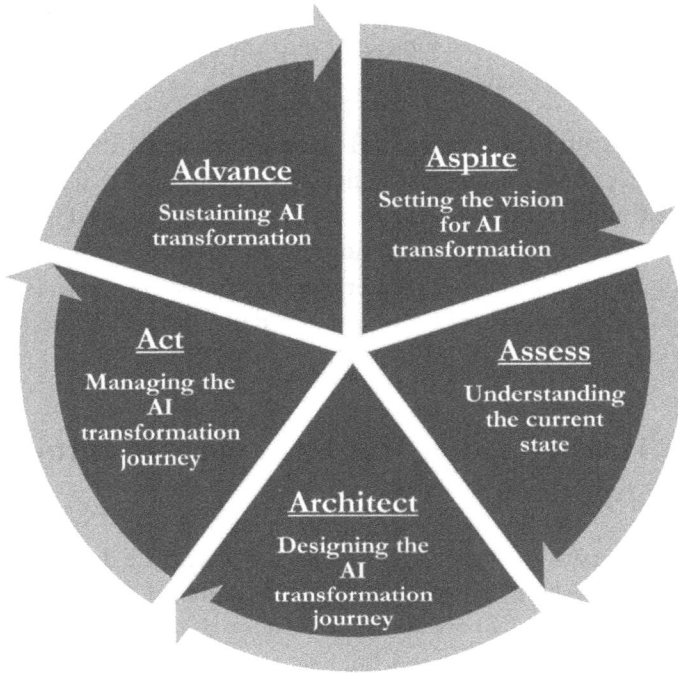

Figure 17-4: The Five Frames of Performance and Health Approach for AI Transformation

Aspire: Setting the Vision for AI Transformation

The "Aspire" stage in transforming to an AI Native Enterprise involves setting a compelling vision and ambitious targets for the integration and utilization of Artificial Intelligence (AI) within the organization. This foundational step is crucial as it sets the direction and tone for the transformation journey, aligning stakeholders around a shared purpose and objectives. Here's how organizations can effectively navigate the "Aspire" stage:

Crafting a Clear AI Vision

- **Define the Role of AI**: Clearly articulate how AI will enhance the organization's capabilities, from improving operational efficiency to innovating products and services. The vision should reflect how AI aligns with the broader business strategy and goals.

- **Inspire Stakeholders**: The AI vision should inspire and excite stakeholders, painting a picture of how AI will transform the organization and its value proposition to customers, employees, and partners.

Setting Ambitious, Yet Achievable Targets

- **Identify Key Performance Indicators (KPIs)**: Establish specific, measurable objectives that AI initiatives aim to achieve. These could include metrics related to customer satisfaction, operational efficiency, revenue growth, or innovation.

- **Balance Ambition with Realism**: While targets should be ambitious to drive significant change, they also need to be achievable to maintain credibility and motivation. Setting milestones for short-term wins alongside long-term goals can help balance this.

Engaging Stakeholders

- **Broad Communication**: Use various platforms and forums to communicate the AI vision and targets to all stakeholders, ensuring the message is clear and consistent across the organization.

- **Build a Coalition of Support**: Identify and engage champions within the organization who can advocate for the AI transformation. This includes leaders at all levels, influential employees, and key external partners.

Fostering Buy-in

- **Highlight the Benefits**: Clearly articulate the benefits of the AI transformation for different stakeholder groups, including how it will address current pain points and create new opportunities.

- **Address Concerns and Questions**: Openly discuss any concerns stakeholders may have about the AI transformation, including job impacts, data privacy, and ethical considerations. Providing clear answers and solutions will help build trust and support.

The "Aspire" stage is critical for successfully launching an AI transformation initiative. By setting a clear vision and ambitious targets, organizations can align their efforts towards a common goal, ensuring that all stakeholders are motivated and moving in the same direction. This stage lays the groundwork for the subsequent steps in the transformation journey, creating a strong foundation for achieving AI integration and innovation.

Assess: Understanding the Current State

The "Assess" stage is a critical step in the journey towards becoming an AI Native Enterprise, requiring organizations to conduct a thorough evaluation of their current capabilities, resources, and environment in relation to AI. This assessment provides a clear understanding of where the organization stands, identifying strengths to leverage, gaps to fill, and challenges to overcome. It lays the groundwork for informed decision-making and strategic planning in the AI transformation process.

Evaluate AI Readiness

- **Technology Infrastructure**: Assess the current state of the organization's technology infrastructure to determine if it can support AI initiatives. This includes data architecture, computing resources, and integration capabilities.

- **Data Quality and Availability**: Evaluate the quality, quantity, and accessibility of data, as AI technologies heavily rely on data to generate insights and drive decisions. Identify any gaps in data collection, storage, and governance practices.

- **Talent and Skills**: Take stock of the organization's talent pool in terms of AI expertise, including data scientists, AI engineers, and other related roles. Assess the need for training or hiring to build AI capabilities.

- **Cultural Readiness**: Gauge the organization's cultural openness to AI and change. This includes employees' willingness to adopt new technologies, innovate, and adapt to new ways of working.

Benchmark Against Best Practices

- **Industry Comparison**: Compare the organization's AI maturity level against industry benchmarks and best practices. This helps in understanding the competitive landscape and identifying areas where the organization needs to improve or can potentially lead.

- **Success Stories and Lessons Learned**: Look at case studies and examples of successful AI implementations within and outside the industry. Analyzing these can provide valuable insights and inspiration for the organization's AI journey.

Identify Opportunities and Challenges

- **AI Integration Opportunities**: Identify specific areas within the organization where AI can add significant value. This could include operational processes that can be optimized, customer experiences that can be enhanced, or new products and services that can be developed using AI.

- **Anticipate Implementation Challenges**: Understand the potential obstacles to AI adoption, including technical challenges, regulatory and compliance issues, ethical considerations, and resistance from employees or other stakeholders.

Conducting a SWOT Analysis

A SWOT (Strengths, Weaknesses, Opportunities, Threats) analysis can be a useful tool at this stage, offering a structured way to evaluate the internal and external factors that could impact the AI transformation journey. This includes:

- **Strengths**: Internal attributes and resources that support successful AI implementation.

- **Weaknesses**: Internal limitations or gaps that could hinder AI initiatives.

- **Opportunities**: External conditions that could favor the adoption and success of AI within the organization.

- **Threats**: External challenges that could pose risks to AI initiatives.

The "Assess" stage is about taking a hard look at the organization's current state and its readiness for AI transformation. By thoroughly evaluating the existing capabilities, resources, and the broader ecosystem, organizations can clearly understand the starting point of their AI journey. This assessment not only highlights the practical steps needed to prepare for AI integration but also aligns expectations and strategies with the organization's actual conditions and capabilities. It ensures that the subsequent steps in the AI transformation process are based on a solid understanding of where the organization stands today and where it needs to go.

Architect: Designing the AI Transformation Journey

The "Architect" stage is pivotal in the transformation to an AI Native Enterprise, where strategic planning translates the vision and insights gained from the assessment phase into a concrete, actionable roadmap. This phase involves designing the initiatives, structures, and processes that will guide the organization from its current state to its desired future state, where AI is seamlessly integrated into its operations and culture.

Develop a Strategic AI Roadmap

- **Define Key Initiatives**: Based on the assessment outcomes, identify specific AI projects and initiatives that align with the organization's strategic objectives and AI vision. This could include developing new AI-based products, automating processes, or enhancing decision-making with AI analytics.

- **Sequence Initiatives**: Prioritize the initiatives based on their potential impact, feasibility, and the organization's readiness. Start with projects that can deliver quick wins or foundational capabilities necessary for more complex AI applications in the future.

Establish Governance and Support Structures

- **AI Governance Framework**: Develop a governance framework that defines roles, responsibilities, and decision-making processes for AI initiatives. This framework should ensure alignment with business goals, ethical standards, and compliance requirements.

- **Cross-Functional AI Teams**: Form cross-functional teams comprising business leaders, AI experts, data scientists, IT professionals, and change management specialists. These teams are tasked with driving the AI initiatives, ensuring they meet business needs and are implemented effectively.

- **Investment in Infrastructure and Talent**: Allocate resources for developing or upgrading technology infrastructure essential for AI, such as cloud computing services, data storage solutions, and advanced analytics tools. Plan for talent acquisition or development programs to build the necessary AI skills within the organization.

Design the Change Management Strategy

- **Stakeholder Engagement Plan**: Create a comprehensive plan to engage stakeholders throughout the AI transformation journey. This includes regular communication, education, and involvement in the change process to build support and address concerns.

- **Cultural Change Initiatives**: Recognize the cultural shifts required to embrace AI and foster a culture of innovation, agility, and continuous learning. Implement initiatives that promote these cultural attributes, such as innovation labs, hackathons, and AI literacy programs.

- **Training and Development Programs**: Design training and development programs to upskill employees in AI-related areas. This not only includes technical skills for IT and data science teams but also broader AI awareness for all employees to understand the role of AI in their work.

Plan for Scalability and Sustainability

- **Scalable AI Solutions**: Ensure that the AI solutions designed can be scaled across the organization. This involves adopting technologies and platforms that support scalability, as well as designing processes that can be replicated or adapted for different parts of the business.

- **Sustainability Measures**: Incorporate sustainability measures to ensure the long-term success of AI initiatives. This includes establishing ongoing performance monitoring, continuous improvement processes, and mechanisms to adapt to changing business needs and technological advancements.

The "Architect" stage is about meticulously designing the blueprint for AI transformation, detailing the what, how, and when of the journey. By developing a strategic AI roadmap, establishing robust governance and support structures, designing a change management strategy, and planning for scalability and sustainability, organizations set themselves up for successful transformation. This thoughtful planning and design phase ensures that AI initiatives are strategically aligned, technically feasible, and culturally embraced, laying a strong foundation for becoming an AI Native Enterprise.

Act: Managing the AI Transformation Journey

The "Act" stage is where strategic plans for AI transformation begin to materialize through concrete actions and initiatives. It's a critical phase in the journey toward becoming an AI Native Enterprise, focusing on the execution of the AI roadmap, managing the change process, and ensuring that the organization navigates the transformation effectively and efficiently. This stage demands agility, resilience, and continuous engagement from all levels of the organization.

Implement AI Initiatives

- **Agile Execution**: Implement AI projects using agile methodologies, which allow for flexibility, rapid iteration, and adjustment based on feedback and outcomes. This approach helps in managing the complexities and uncertainties inherent in AI projects.

- **Project Management and Oversight**: Establish strong project management practices to ensure AI initiatives stay on track, within budget, and aligned with business objectives. Regular status updates, risk management, and stakeholder engagement are crucial components.

- **Pilot Programs and Scaling**: Start with pilot programs to test AI solutions in controlled environments before scaling them across the organization. Pilots help in identifying potential issues, assessing the impact, and making necessary adjustments before wider implementation.

Foster an AI-Enabled Culture

- **Promote AI Literacy**: Continuously educate and inform the workforce about AI and its potential impact on their work and the organization. This includes training programs, workshops, and regular communications that demystify AI and encourage a positive outlook towards its adoption.

- **Encourage Collaboration and Innovation**: Create an environment that encourages experimentation, collaboration, and innovation. Encourage teams to share ideas, learn from failures, and celebrate successes in AI projects.

- **Address Resistance and Concerns**: Actively address any resistance or concerns from employees regarding AI adoption. This involves transparent communication about the role of AI, its benefits, and how it will affect their work, as well as providing support and training to ease the transition.

Manage Change and Transition

- **Change Leadership**: Engage leaders at all levels to champion the AI transformation, demonstrating commitment, and leading by example. Leaders should be visible advocates of change, addressing challenges head-on and motivating their teams throughout the transformation journey.

- **Stakeholder Engagement and Communication**: Maintain open lines of communication with all stakeholders, providing regular updates on progress, celebrating milestones, and discussing challenges. Tailor communication strategies to different stakeholder groups to ensure relevance and effectiveness.

- **Support Structures and Resources**: Provide adequate support structures, such as help desks, training sessions, and mentorship programs, to assist employees in adapting to new AI-enabled processes and tools.

Monitor, Evaluate, and Adjust

- **Performance Metrics and KPIs**: Establish clear metrics and KPIs to measure the success of AI initiatives against the objectives outlined in the AI roadmap. This includes both quantitative measures, such as productivity improvements and cost savings, and qualitative measures, such as employee satisfaction and customer experience.

- **Feedback Loops**: Implement mechanisms to gather feedback from employees, customers, and other stakeholders on AI projects. Use this feedback to make continuous improvements and adjustments to AI initiatives.

- **Iterative Improvement**: Treat the AI transformation as an ongoing process rather than a one-time project. Continuously assess the performance of AI implementations, learn from experiences, and iterate to enhance and expand AI capabilities across the organization.

The "Act" stage is where the strategic vision for AI transformation begins to take shape through practical implementation. It requires a balanced approach of disciplined execution, cultural enablement, and proactive change management. By effectively managing the AI transformation journey, organizations can ensure that they not only achieve their AI objectives but also build a resilient, agile, and innovative culture that can sustain and enhance AI capabilities over time.

Advance: Sustaining AI Transformation

The "Advance" stage is crucial for ensuring that the momentum gained during the AI transformation journey is not only maintained but also built upon. This phase focuses on embedding AI deeply into the organizational fabric, making it a continuous driver of innovation, efficiency, and competitive advantage. Sustaining AI transformation requires ongoing commitment, strategic foresight, and an adaptive approach to dealing with evolving technologies and market dynamics.

Institutionalize AI Learning

- **Continuous Education and Training**: Establish continuous learning programs to keep the workforce updated on the latest AI technologies, methodologies, and best practices. This could include in-house training, online courses, workshops, and participation in AI conferences and seminars.

- **Knowledge Sharing Platforms**: Develop platforms or forums where employees can share insights, challenges, successes, and learnings related to AI projects. This fosters a culture of knowledge sharing and collective problem-solving.

Leverage AI for Innovation

- **Innovation Labs and Incubators**: Set up dedicated spaces or programs like innovation labs or incubators to explore new AI technologies and applications. These units can work on speculative projects with the potential to create new value streams or significantly enhance operational efficiencies.

- **Crowdsourcing Ideas**: Engage the broader organization in ideation processes to identify new areas where AI can be applied. Crowdsourcing can uncover innovative applications of AI from employees who are deeply familiar with the challenges and opportunities in their daily work.

Embed Continuous Improvement Processes

- **Iterative Development**: Adopt an iterative approach to AI project development, where projects are continuously refined and improved based on feedback and performance data. This ensures that AI solutions remain aligned with business needs and can adapt to changes quickly.

- **AI Performance Monitoring**: Implement robust monitoring systems to track the performance of AI initiatives against defined KPIs. Use data-driven insights to make informed decisions about scaling, optimizing, or pivoting AI projects.

Establish AI Governance and Ethical Standards

- **AI Governance Frameworks**: Develop and maintain comprehensive AI governance frameworks that ensure AI projects align with organizational goals, comply with regulations, and adhere to ethical standards. This includes oversight of data usage, model transparency, and fairness.

- **Ethical AI Use**: Reinforce the importance of ethical considerations in AI development and deployment. Establish principles and guidelines for ethical AI that address issues such as bias, privacy, and transparency.

Foster an AI-Adaptive Organizational Culture

- **Promote an AI-First Mindset**: Encourage leaders and employees to think "AI-first" when considering solutions to business challenges, embedding AI as a natural part of problem-solving and innovation processes.

- **Adaptability and Resilience**: Cultivate an organizational culture that values adaptability and resilience, recognizing that the AI landscape is constantly evolving. Prepare the organization to pivot as new AI technologies and applications emerge.

The "Advance" stage is about ensuring that AI transformation becomes a continuous journey rather than a destination. It requires embedding AI into the organization's DNA, fostering a culture of continuous learning and innovation, and establishing robust governance and ethical frameworks. By focusing on sustaining and advancing AI capabilities, organizations can not only reap immediate benefits but also position themselves to capture future opportunities, maintaining a competitive edge in an increasingly AI-driven world.

Conclusion

The transition to an AI Native Enterprise is an intricate journey that demands a strategic and holistic approach, touching upon both the technological advancements and the cultural shifts within an organization. By adopting the Five Frames of Performance and Health, organizations are equipped to adeptly manage the complexities of AI integration. This framework not only promises enhanced performance through innovation and operational efficiency but also cultivates a resilient and agile culture, essential for thriving in the AI era. This transformative path requires unwavering commitment, visionary leadership, and an organization-wide openness to embrace change. It's a journey that, when navigated successfully, leads to enduring success, positioning organizations to flourish in the dynamic and continuously evolving digital landscape.

Strategic Paths to AI Transformation: A Comparative Guide

In the realm of organizational change, particularly with the integration of Artificial Intelligence (AI), four transformation approaches stand out: Business-Driven AI-Native Transformation, Dual Transformation, Dual Operating System, and Five Frames of Performance and Health. Each method offers a unique strategy for steering organizations through the complexities of significant change. This section provides a comparative analysis of these frameworks, highlighting their characteristics and how they align with companies of varying sizes.

Approach	Dual Transformation	Dual Operating System	Five Frames of Performance and Health	Business-Driven AI-Native Transformation
Foundation Concept	Separation into two paths: optimizing the core while exploring new innovations	Parallel hierarchical and network systems for efficiency and innovation.	Sequential process focusing on both performance outcomes and organizational health	Phased approach aligning AI integration with strategic business transformation
Primary Focus	Balancing optimization and innovation for growth	Enhancing agility without disrupting core efficiency.	Comprehensive change management, addressing performance and health	Strategic business transformation through AI, from process augmentation to business reinvention
Key Components	Two transformations linked by shared capabilities	Hierarchical for stability, network for agility	Aspire, Assess, Architect, Act, Advance	Business Augmentation, Business Transformation, Business Reinvention
Execution Complexity	Moderate to High	High	Moderate to High	Low to High
Change Management Focus	High	Moderate	High	High
Suitability for Company Size	Large	Medium to Large	Small to Large	Small to Large

Table 17-1: The Comparison of Four AI Transformation Approaches

Suitability for Company Size

Small Companies: The Five Frames of Performance and Health is an excellent match for small companies, offering a structured yet adaptable framework that aligns with their agile environments and resource limitations. The Business-Driven AI-Native Transformation framework also stands out as particularly suitable for small enterprises. Its early focus on Business Augmentation allows for immediate operational enhancements through AI. Moreover, its strategic design supports small companies in pursuing more ambitious business reinventions in a nimble and rapid manner, leveraging their inherent agility to adapt and innovate quickly. This combination of frameworks provides small businesses with a solid foundation for AI integration, enabling them to streamline operations and potentially leapfrog into advanced stages of business transformation with greater speed and flexibility.

Medium Companies: Medium-sized companies find a strategic ally in the Business-Driven AI-Native Transformation framework, which supports a comprehensive journey from enhancing current operations to complete business model reinvention. The Dual Transformation approach also resonates with medium enterprises, enabling them to balance optimization of core processes with innovation, provided they have the necessary resources.

Large Companies: For large organizations, the Dual Operating System approach is particularly advantageous, catering to the need for sustaining two parallel structures that foster innovation while maintaining core operations. Large enterprises can also significantly benefit from the holistic and comprehensive approach of the Business-Driven AI-Native Transformation, aiming for market differentiation and industry leadership through AI.

Choosing the Right Approach

Selecting the ideal approach for AI transformation hinges on a blend of factors, including an organization's scale, composition, available resources, and precise objectives for embedding AI and undergoing digital transformation. The Dual Transformation and Dual Operating System methods present cutting-edge frameworks designed to harmonize core operational efficiency with agility and innovation. However, their successful deployment demands substantial resources and a deep commitment from management, rendering them more compatible with larger enterprises. Conversely, the Five Frames of Performance and Health model offers a versatile framework that accentuates both the tangible outcomes and the underlying organizational vitality necessary for enduring change, making it suitable for a broad range of company sizes and developmental phases.

The newly introduced Business-Driven AI-Native Transformation Framework enriches this landscape by offering a sequential, strategic roadmap for AI integration, appealing to enterprises across the spectrum seeking thorough transformation. Each framework brings distinct benefits, shaped by varying organizational dimensions, structures, and strategic aims. Choosing the most appropriate transformation pathway necessitates a deliberate evaluation of an organization's particular demands, capabilities, and aspirations for AI incorporation.

Closing: Embarking on the AI Native Enterprise Journey

As we conclude this final chapter and the book, "Transforming to an AI Native Enterprise," we stand on the precipice of a new era in business and technology. The journey through the myriad landscapes of AI transformation has equipped us with insights, strategies, and frameworks essential for navigating the complexities of becoming an AI Native Enterprise. This chapter has not only served as a capstone to our exploration but also as a launchpad for organizations ready to embark on this transformative journey.

The road to becoming an AI Native Enterprise is marked by continuous learning, adaptation, and strategic foresight. It requires a deep commitment to not just adopting AI technologies but embedding them into the very DNA of the organization. This transformation transcends operational efficiency and product innovation; it's about cultivating a culture of innovation, agility, and resilience that can thrive in the AI-driven future.

As we close this chapter and the book, it's important to recognize that the journey does not end here. The landscape of AI and digital transformation is ever-evolving, with new technologies, methodologies, and market dynamics emerging at a rapid pace. Staying ahead in this dynamic environment requires an ongoing commitment to innovation, a willingness to experiment and learn from failures, and the agility to pivot strategies in response to new opportunities and challenges.

The frameworks and approaches discussed in this chapter provide a roadmap for initiating and sustaining AI transformation efforts. Whether you choose the Dual Transformation approach to balance core optimization with innovation, the Dual Operating System model for enhanced agility, the Five Frames of Performance and Health for a comprehensive change management strategy, or the Business-Driven AI-Native Transformation Framework for a phased strategic transformation, the key is to align AI initiatives with your organization's strategic objectives and cultural dynamics.

As this is the last chapter of our book, we hope it serves not only as a culmination of the insights gathered throughout but also as an inspiration for leaders, innovators, and change-makers to take bold steps toward integrating AI into their enterprises. The journey to becoming an AI Native Enterprise is both challenging and rewarding, offering unparalleled opportunities for growth, innovation, and leadership in the digital age.

Let this chapter be a beacon, guiding your organization as it navigates the complexities of AI transformation, and may your journey be marked by success, innovation, and the realization of your strategic vision. The future belongs to those who dare to embrace AI, not just as a technological tool, but as a cornerstone of their enterprise strategy. Welcome to the dawn of your AI Native Enterprise.

Acknowledgements

Embarking on the journey of writing this book has been nothing short of an adventure, demanding not just courage and perseverance, but also the collective support of a remarkable community. Today, as I stand on the verge of its completion, my heart overflows with gratitude for the myriad of souls and serendipities that have steered this voyage.

To my cherished colleagues, your relentless encouragement and shared passion for Generative AI and digital transformation have been the bedrock of this book's inception. Each discussion, argument, and brainstorming session infused our collective narrative with invaluable insights, enriching it beyond measure.

My profound thanks go to my CIO and CTO peers, whose rich dialogues and shared experiences underscored the imperative of disseminating my journey and expertise in AI-powered business transformation. Your perspectives have not only shaped my AI-Native thinking but also underscored the significance of contributing our knowledge to the wider community, inspiring both current and future AI aficionados.

I am immensely grateful to the legion of AI scholars and practitioners whose pioneering contributions to artificial intelligence have been nothing short of inspirational. Your extensive research and deep dives into Generative AI and its transformative potential have profoundly influenced my perspective, enriching my AI-Native approach to both thought and application.

In this digital epoch, the wonders of AI tools like ChatGPT, Google Bard, and others have been indispensable. Their utility in boosting my productivity, managing my dense schedule, and simplifying complex tasks has been invaluable. My deepest appreciation goes to these digital aides and their creators for their pivotal role in my journey.

Above all, my heart brims with gratitude for my family. To my wife, Yan Chen, and our son, Henry—your infinite love, patience, and unwavering support have been the cornerstone of this dream. Amidst the whirlwind of life's commitments and hurdles, your steadfast belief and love have been my sanctuary.

To each one of you who has touched this journey, my heartfelt thanks. Your support has transformed this book from a mere idea into a beacon of shared knowledge and inspiration.

About the Author

Yi Zhou is a globally acclaimed AI thought leader and executive, renowned for his pioneering work in AI-powered business transformation. Serving as the Chief Technology Officer (CTO) and Chief Information Officer (CIO) at Adaptive Biotechnologies, Yi has a remarkable history of leadership at the nexus of technology and healthcare, including transformative roles at GE Healthcare and Quest Diagnostics. He is celebrated for his visionary work, notably establishing the GE Healthcare AI Standard and Playbook and leading the launch of the world's first AI-driven, FDA-approved medical imaging devices, including X-ray and MRI technologies. He played a crucial role in setting industry benchmarks through his involvement in the AI Committee of the Medical Imaging & Technology Alliance (MITA).

Beyond his achievements in technology and leadership, Yi is a prolific author. He co-authored the widely acknowledged O'Reilly book "97 Things Every Software Architect Should Know" and has made significant publications across AI, software architecture, cybersecurity, and life sciences. Notably, Yi authored the bestselling Generative AI book "Prompt Design Patterns: Mastering the Art and Science of Prompt Engineering", further underscoring his status as a leading voice in the AI community.

Yi Zhou's accolades are extensive, showcasing his leadership and innovation in the field. He has been recognized as a leading executive in American Healthcare Leader magazine and was a finalist for the "CIO of the Year" 2023 Seattle ORBIE Award. Additionally, Yi has been honored with multiple CEO and DNA awards, highlighting his visionary approach and commitment to innovation. Beyond these honors, his dedication extends to education and mentorship, serving as a board member at the University of Washington Information School and advising over 50 startups and investment firms, illustrating his significant influence on both the academic and entrepreneurial landscapes.

Yi holds dual master's degrees in Computer Science and Microbiology, demonstrating his diverse expertise. Yi holds dual master's degrees in Computer Science from the University of Missouri and Microbiology from the University of Kansas Medical Center, demonstrating his diverse expertise. A bachelor's degree in Microbiology from the Fudan University and four professional certificates in Agile process and software development further underscore his dedication to continuous learning and mastery in his domain.

References and Further Reading

1. Ajgaonkar, Amol. "MLOps: The Key to Unlocking AI Operationalization." Insight.com, 2021. https://www.insight.com/en_US/content-and-resources/tech-journal/summer-2021/mlops-the-key-to-unlocking-ai-operationalization.html

2. AlgorithmWatch, "AI Ethics Guidelines Global Inventory." 2020. https://algorithmwatch.org/en/ai-ethics-guidelines-global-inventory/

3. Apostolopoulos, Ioannis D.; Tzani, Mpesi; Aznaouridis, Sokratis I. "Chat-GPT: ascertaining the self-evident. The use of AI in generating human knowledge." arXiv:2308.06373, 2023.

4. Barnes, C.; Lee, J. "Cultural Transformation for AI Adoption in Organizations." Organizational Dynamics, 33(1), 12-29, 2024.

5. Barrett, Clark, et al. "Identifying and Mitigating the Security Risks of Generative AI." arXiv:2308.14840, 2023.

6. Bommasani, Rishi, et al. "On the Opportunities and Risks of Foundation Models." arXiv:2108.07258, 2022.

7. Bubeck, Sébastien, et al. "Sparks of Artificial General Intelligence: Early experiments with GPT-4." arXiv:2303.12712, 2023.

8. Cameron, Esther; Green, Mike. "Making Sense of Change Management: A Complete Guide to the Models, Tools and Techniques of Organizational Change." 5th Edition, Kogan Page, 2019.

9. Chandrasekaran, Arun; Miclaus, Radu; Goodness, Eric. "A CTO's Guide to the Generative AI Technology Landscape." Gartner, 2023.

10. Choo, Julie; Christison, Graham; Adnan, Dana. "THE STRATEGY JOURNEY: How to transform your business operating model in the digital age with value-driven, customer co-created and network-connected services." Stratability Academy, 2020.

11. Choudhary, Farhan; Sicular, Svetlana. "A Comprehensive Guide to Responsible AI." Gartner, 2022.

12. Coyle, Daniel. "The Culture Code: The Secrets of Highly Successful Groups." Bantam, 2018.

13. Czerwonko, Alejo; White, Jules. "Life after the hype: How AI is transforming industries and economies." World Economic Forum, 2023. https://www.weforum.org/agenda/2023/12/life-after-the-hype-how-ai-is-transforming-industries-and-economies/

14. Daugherty, P. R.;Wilson, H. J. "Human + Machine: Reimagining Work in the Age of AI." Harvard Business Review Press, 2018.

15. Davenport, Thomas H.; Mittal, Nitin. "All-in On AI: How Smart Companies Win Big with Artificial Intelligence." Harvard Business Review Press, 2023.

16. Davenport, T. H.; Ronanki, R. "Artificial Intelligence for the Real World." Harvard Business Review, 2018.

17. Davidson, H.; Zhao, F. "AI Integration in Global Enterprises: Challenges and Successes." Technology Management Journal, 17(3), 78-94, 2023.

18. Deloitte, "Building Trustworthy Generative AI." 2023. https://www2.deloitte.com/content/dam/Deloitte/us/Documents/consulting/us-ai-institute-trusted-generative-ai.pdf

19. Deloitte, "Now decides next: Insights from the leading edge of generative AI adoption." 2024. https://www2.deloitte.com/content/dam/Deloitte/us/Documents/consulting/us-state-of-gen-ai-report.pdf

20. Deloitte, "Proactive risk management in Generative AI." 2023. https://www2.deloitte.com/content/dam/Deloitte/us/Documents/deloitte-analytics/us-ai-institute-responsible-use-of-generative-ai.pdf

21. Deloitte, "The Generative AI Dossier: A selection of high-impact use cases across six major industries."

2023. https://www2.deloitte.com/content/dam/Deloitte/us/Documen
ts/consulting/us-ai-institute-gen-ai-use-cases.pdf

22. Dell'Acqua, Fabrizio, et al. "Navigating the Jagged Technological Frontier: Field Experimental Evidence of the Effects of AI on Knowledge Worker Productivity and Quality." Harvard Business School, 2023. https://www.hbs.edu/ris/Publication%20Files/24-013_d9b45b6 8-9e74-42d6-a1c6-c72fb70c7282.pdf

23. Dou, Yao; Laban, Philippe; Gardent, Claire; Xu, Wei. "Automatic and Human-AI Interactive Text Generation." arXiv:2310.03878, 2023.

24. Eloundou, Tyna, et al. "GPTs are GPTs: An Early Look at the Labor Market Impact Potential of Large Language Models." arXiv:2303.10130, 2023.

25. Epstein, Ziv, et al. "Art and the science of generative AI: A deeper dive." arXiv:2306.04141, 2023.

26. Fang, Xiao; Che, Shangkun; Mao, Minjia; Zhang, Hongzhe; Zhao, Ming; Zhao, Xiaohang. "Bias of AI-Generated Content: An Examination of News Produced by Large Language Models." arXiv:2309.09825, 2023.

27. Fountaine, T.; McCarthy, B.; Saleh, T. "Building the AI-Powered Organization." Harvard Business Review, 2019.

28. Gan, Wensheng; Wan, Shicheng; Yu, Philip S. "Model-as-a-Service (MaaS): A Survey." arXiv:2311.05804, 2023.

29. Gao, Yunfan, et al. "Retrieval-Augmented Generation for Large Language Models: A Survey." arXiv:2312.10997, 2024.

30. Goertzel, Ben. "Generative AI vs. AGI: The Cognitive Strengths and Weaknesses of Modern LLMs." arXiv:2309.10371, 2023.

31. Goodfellow, I.; Bengio, Y.; Courville, A.; Bengio, Y. "Deep Learning." MIT Press, 2016.

32. Gozalo-Brizuela, Roberto; Garrido-Merchán, Eduardo C. "A survey of Generative AI Applications." arXiv:2306.02781, 2023.

33. Greenfield, S. "Why Transformations Fail: A Deep Dive into Organizational Change." Global Business Review, 21(4), 234-256, 2023.

34. Gupta, Maanak; Akiri, CharanKumar; Aryal, Kshitiz; Parker, Eli; Praharaj, Lopamudra. "From ChatGPT to ThreatGPT: Impact of Generative AI in Cybersecurity and Privacy." arXiv:2307.00691, 2023.

35. Hashimoto, Y. "Towards a More Agile AI: Implementing Agile Practices in AI Development Teams." Journal of AI and Technology, 2021.

36. Holley, Kerrie; Becker, Siupo, M.D. "AI-First Healthcare: AI Applications in the Business and Clinical Management of Health." O'Reilly Media, 2021.

37. Huschens, Martin; Briesch, Martin; Sobania, Dominik; Rothlauf, Franz. "Do You Trust ChatGPT? -- Perceived Credibility of Human and AI-Generated Content." arXiv:2309.02524, 2023.

38. Iansiti, Marco; Lakhani, Karim R. "Competing in the Age of AI: Strategy and Leadership When Algorithms and Networks Run the World." Harvard Business Review Press, 2020.

39. Jackson, Thomas L. "Hoshin Kanri for the Lean Enterprise: Developing Competitive Capabilities and Managing Profit." Productivity Press, 2006.

40. Janiak, Stacy. "3 Ways To Harness GenAI As A Business Disruptor." Forbes, 2024. https://www.forbes.com/sites/deloitte/2024/01/17/3-ways-to-harness-genai-as-a-business-disruptor/?sh=24bf39615e6e

41. Jobin, A.; Ienca, M.; Vayena, E. "The global landscape of AI ethics guidelines." Nature Machine Intelligence, 2019.

42. Johnston, Kristin. "Five compliance best practices for a successful AI governance program." Iapp.org, 2023. https://iapp.org/news/a/five-compliance-best-practices-for-a-successful-ai-governance-program/

43. Johnson, L.; Kumar, R. "The AI-Powered Enterprise: From Automation to Reinvention." International Review of Business Research, 15(2), 89-115, 2024.

44. Jurcys, Paulius; Fenwick, Mark. "Originality and the Future of Copyright in an Age of Generative AI." arXiv:2309.13055, 2023.

45. Kaur, Jagreet. "The Complete Guide to Generative AI Architecture." Xenonstack.com, 2023. https://www.xenonstack.com/blog/generative-ai-architecture

46. Keller, Scott; Schaninger, Bill. "Beyond Performance 2.0: A Proven Approach to Leading Large-Scale Change." 2nd edition. Wiley, 2019.

47. Kotter, John P. "Accelerate: Building Strategic Agility for a Faster-Moving World." Harvard Business Review Press, 2014.

48. Lamarre, Eric; Samaje, Kate; Zemmel, Rodney. "Rewired: The McKinsey Guide to Outcompeting in the Age of Digital and AI." Wiley, 2023.

49. Lafley, A.G.; Martin, Roger L. "Playing to Win: How Strategy Really Works." Harvard Business Review Press, 2013.

50. Lee, K-F. "AI Superpowers: China, Silicon Valley, and the New World Order." Houghton Mifflin Harcourt, 2018.

51. Leonardi, Paul; Neeley, Tsedal. "The Digital Mindset: What It Really Takes to Thrive in the Age of Data, Algorithms, and AI." Harvard Business Review Press, 2022.

52. Loukides, Mike. "Generative AI in the Enterprise." O'Reilly, 2023. https ://www.oreilly.com/radar/generative-ai-in-the-enterprise/

53. Lu, Qinghua, et al. "Building the Future of Responsible AI: A Reference Architecture for Designing Large Language Model based Agents." arXiv :2311.13148, 2023.

54. Mahmood, Razi; Wang, Ge; Kalra, Mannudeep; Yan, Pingkun. "Fact-Checking of AI-Generated Reports." arXiv:2307.14634, 2023.

55. Mainali, Mallika; Weber, Rosina O. "What's meant by explainable model: A Scoping Review." arXiv:2307.09673, 2023.

56. Mariani, Joe; Eggers, William D.; Kishnani, P. K. "The AI regulations that aren't being talked about: Patterns in AI policies can expose new opportunities for governments to steer AI's development." Deloitte. https://www2.deloitte.com/us/en/insights/industry/public-secto r/ai-regulations-around-the-world.html

57. McAfee, A.; Brynjolfsson, E. "Machine, Platform, Crowd: Harnessing Our Digital Future." W. W. Norton & Company, 2017.

58. McKinsey, "Exploring opportunities in the generative AI value chain." 2023. https://www.mckinsey.com/capabilities/quantumblack/our-insig hts/exploring-opportunities-in-the-generative-ai-value-chain

59. McKinsey, "Generative AI and the future of work in America." 2023. https://www.mckinsey.com/mgi/our-research/generative-ai-and-the-future-of-work-in-america

60. McKinsey, "Leading your organization to responsible AI." 2019. https://www.mckinsey.com/capabilities/quantumblack/our-insights/leading-your-organization-to-responsible-ai

61. McKinsey, "The economic potential of generative AI: The next productivity frontier." 2023. https://www.mckinsey.com/capabilities/mckinsey-digital/our-insights/the-economic-potential-of-generative-ai-the-next-productivity-frontier

62. McKinsey, "The next-generation operating model for the digital world." 2017. https://www.mckinsey.com/capabilities/mckinsey-digital/our-insights/the-next-generation-operating-model-for-the-digital-world

63. McKinsey, "The state of AI in 2023: Generative AI's breakout year." 2023. https://www.mckinsey.com/capabilities/quantumblack/our-insights/the-state-of-ai-in-2023-generative-ais-breakout-year

64. McIlwain, Matt. "GenAI in 2024 – Another Decade in One Year?" Madrona, 2023. https://www.madrona.com/genai-in-2024-another-decade-in-one-year/?utm_source=Newsletter&utm_medium=Social&utm_campaign=GenAI+in+2024

65. Mohamadi, Salman; Mujtaba, Ghulam; Le, Ngan; Doretto, Gianfranco; Adjeroh, Donald A. "ChatGPT in the Age of Generative AI and Large Language Models: A Concise Survey." arXiv:2307.04251, 2023.

66. Nerbun, J.P. "The Culture System: A Proven Process for Creating an Extraordinary Team Culture." TOC Culture Consulting, 2022.

67. O'Neill, T.; Singh, G. "Strategic AI Deployment in Business Operations." Strategic Business Review, 9(2), 150-175, 2023.

68. Patel, M.; Chang, Y. Adapting to AI: Frameworks for Corporate Evolution. AI & Society, 29(1), 101-122, 2024.

69. Poitevin, Helen. "What Generative AI Means for Your Talent Strategy." Gartner, 2023.

70. Raj, P.; Raman, A. "Feature Stores for Machine Learning." ACM Computing Surveys, 2020.

71. Ramos, Leinar, et al. "How to Pilot Generative AI." Gartner, 2023.

72. Renieris, Elizabeth M.; Kiron, David; Mills, Steven. "Building Robust RAI Programs as Third-Party AI Tools Proliferate." MIT Sloan Management Review, 2023.

73. Schwartz, J.; Bonnet, D.; Anderson, M. "The Technology Fallacy: How People Are the Real Key to Digital Transformation." MIT Press, 2020.

74. Shokrollahi, Yasin; Yarmohammadtoosky, Sahar; Nikahd, Matthew M.; Dong, Pengfei; Li, Xianqi; Gu, Linxia. "A Comprehensive Review of Generative AI in Healthcare." arXiv:2310.00795, 2023.

75. Simons, Adam. "Generative AI Landscape: Applications, Models, Infrastructure." Dgtl Infra, 2023.

76. Smith, J.; Doe, A. "Navigating the AI Revolution: Strategies for Success." Journal of Business Innovation, 12(3), 45-67, 2023.

77. Somers, Meredith. "How generative AI can boost highly skilled workers' productivity." MIT Management Sloan School, 2023. https://mitsloan.mit.edu/ideas-made-to-matter/how-generative-ai-can-boost-highly-skilled-workers-productivity

78. Sweet, Julie, et al. "Total Enterprise Reinvention: The strategy that leads to a new performance frontier." Accenture, 2023.

79. Taff Inc. "Top 10 MLOps Predictions and Trends to Look Out For in 2024." 2024.

80. Tetu, Louis. "Mastering the power of generative AI: Ask these eight questions before your enterprise takes the leap." FastCompany.com, 2023. https://www.fastcompany.com/90956852/mastering-the-power-of-generative-ai-ask-these-eight-questions-before-your-enterprise-takes-the-leap

81. Treacy, Michael; Wiersema, Fred. "The Discipline of Market Leaders: Choose Your Customers, Narrow Your Focus, Dominate Your Market." Basic Books, 1995.

82. Treveil, M.; Laney, D.; Fawcett, C.; Hammersley, J. "Introducing MLOps." O'Reilly Media, 2020.

83. Tung, Teresa. "7 architecture considerations for generative AI." Accenture, 2023. https://www.accenture.com/us-en/blogs/cloud-computing/7-generative-ai-architecture-considerations

84. Vidrih, Marko; Mayahi, Shiva. "Generative AI-Driven Storytelling: A New Era for Marketing." arXiv:2309.09048, 2023.

85. Wang, Haonan, et al. "Can AI Be as Creative as Humans?" arXiv:2401.01623, 2024.

86. Wang, Yuntao; Pan, Yanghe; Yan, Miao; Su, Zhou; Luan, Tom H. "A Survey on ChatGPT: AI-Generated Contents, Challenges, and Solutions." arXiv:2305.18339, 2023.

87. Woodruff, Allison, et al. "How Knowledge Workers Think Generative AI Will (Not) Transform Their Industries." arXiv:2310.06778, 2023.

88. Wu, Jiayang, et al. "AI-Generated Content (AIGC): A Survey." arXiv:2304.06632, 2023.

89. Xia, Boming, et al. "From Principles to Practice: An Accountability Metrics Catalogue for Managing AI Risks." arXiv:2311.13158, 2024.

90. Zhang, Dong. "Should ChatGPT and Bard Share Revenue with Their Data Providers? A New Business Model for the AI Era." arXiv:2305.02555, 2023.

91. Zhou, Yi. "Prompt Design Patterns: Mastering the Art and Science of Prompt Engineering." ArgoLong Publishing, 2023.